Essential Biology for Cambridge IGCSE®
Workbook

For the updated syllabus

Ron Pickering

Oxford excellence for Cambridge IGCSE®

OXFORD

Great Clarendon Street, Oxford, OX2 6DP, United Kingdom

Oxford University Press is a department of the University of Oxford. It furthers the University's objective of excellence in research, scholarship, and education by publishing worldwide. Oxford is a registered trade mark of Oxford University Press in the UK and in certain other countries

© Oxford University Press 2016

The moral rights of the authors have been asserted

First published in 2016

All rights reserved. No part of this publication may be reproduced, stored in a retrieval system, or transmitted, in any form or by any means, without the prior permission in writing of Oxford University Press, or as expressly permitted by law, by licence or under terms agreed with the appropriate reprographics rights organization. Enquiries concerning reproduction outside the scope of the above should be sent to the Rights Department, Oxford University Press, at the address above.

You must not circulate this work in any other form and you must impose this same condition on any acquirer

British Library Cataloguing in Publication Data
Data available

978-0-19-837467-1

1 3 5 7 9 10 8 6 4 2

Paper used in the production of this book is a natural, recyclable product made from wood grown in sustainable forests. The manufacturing process conforms to the environmental regulations of the country of origin.

Printed in China by Golden Cup

Acknowledgements

®IGCSE is the registered trademark of Cambridge International Examinations.

The publishers would like to thank the following for permissions to use their photographs:

Cover image: Corbis Images;

p18:Jiri Hera/Shutterstock; p124:Ron Pickering; p95:Ed Reschke/Getty Images; p107:Design Pics Inc/Alamy Stock Photo.

Artwork by Q2A Media Services Pvt. Ltd. and OUP.

Although we have made every effort to trace and contact all copyright holders before publication this has not been possible in all cases. If notified, the publisher will rectify any errors or omissions at the earliest opportunity.

Links to third party websites are provided by Oxford in good faith and for information only. Oxford disclaims any responsibility for the materials contained in any third party website referenced in this work.

Introduction

When using this workbook you will have the opportunity to develop the knowledge and skills that you need to do well in each of the papers in your IGCSE Biology examination.

The IGCSE syllabus explains that you will be tested in three different ways. These are called Assessment Objectives (AO for short). What these AOs mean to you in the examination is explained below:

Assessment Objective	What the syllabus calls these objectives	What this means in the examination
AO1	Knowledge with understanding	Questions which mainly test your recall (and understanding) of what you have learned. About 50% of the marks in the examination are for AO1.
AO2	Handling information and problem solving	Using what you have learned in unfamiliar situations. These questions often ask you to examine data in tables or graphs, or to carry out calculations. About 30% of the marks are for AO2.
AO3	Experimental skills and investigations	These are tested on the Practical Paper or the Alternative to Practical (20% of the total marks). However, the skills you develop in practising for these papers may well be valuable in handling questions on the theory papers.

Notice that the **recall** questions (AO1) only account for 50% of the marks – you need to show your skill in using these facts for the remaining 50% of the marks.

This workbook contains many exercises to help you to check your recall and to practise these skills. They will be similar to many of the questions you will actually see in your examination, so you will also be helped to develop the skill of working in an examination. In particular, you will find that many of the exercises cover factual material from different parts of the syllabus – exactly like the more difficult questions in the examination. Each worksheet contains a language lab exercise to help you understand and remember key words. There is also a glossary of important terminology at the end of the book.

The answers to the questions are provided, so that you can assess your own performance. Be honest with yourself when checking the marks – you must not be more generous than an examiner would be! Your teacher will probably be able to help you to compare your performance with the expected standards.

Practice may not make perfect, but it will certainly make better.

Good luck!

Ron Pickering

Contents

1 Characteristics and classification of living organisms

1.1	Characteristics of living organisms	2
1.2	Classification	3
1.3	Features of organisms	4
1.4	Vertebrates	5
1.5	Invertebrates	6
1.6	Ferns and flowering plants	7
1.7	Dichotomous keys	8

2 Cells

2.1	Structure of cells	9
2.2	Cell organelles	10
2.3	Different types of cell	11
2.4	Levels of organisation	12

3 Movement in and out of cells

3.1	Diffusion	13
3.2	Osmosis	14
3.3	Osmosis in plant and animal cells	15
3.4	Active transport	16

4 Biological molecules

4.1	Biological molecules	17
4.2	Chemical tests for biological molecules	18
4.3	DNA	19

5 Enzymes

5.1	Structure and action of enzymes	20
5.2	Factors affecting enzyme action: temperature	21
5.3	Enzymes and pH	22

6 Plant nutrition

6.1	Photosynthesis	23
6.2	What is needed for photosynthesis?	24
6.3	Products of photosynthesis	25
6.4	Rate of photosynthesis	26
6.5	Glasshouse production	27
6.6	Leaves	28
6.7	Mineral requirements	29

7 Animal nutrition

7.1	A balanced diet	30
7.2	Sources of nutrients	31
7.3	Balancing energy needs	32
7.4	Starvation and nutrient deficiency	33
7.5	Digestion	34
7.6	Teeth	35
7.7	Mouth, oesophagus, and stomach	36
7.8	Small intestine and absorption	37
7.9	Large intestine and intestinal disease	38

8 Plant transport

8.1	Transport systems	39
8.2	Water uptake	40
8.3	Transpiration	41
8.4	Translocation	42

9 Transport in humans

9.1	Circulation	43
9.2	The heart	44
9.3	Heart and exercise	45
9.4	Blood vessels	46
9.5	Coronary heart disease (CHD)	47
9.6	Blood	48
9.7	Blood in defence	49
9.8	Lymph and tissue fluid	50

10 Diseases and immunity

10.1	Disease	51
10.2	Defence against disease	52
10.3	Aspects of immunity	53
10.4	Controlling the spread of disease	54

11 Gas exchange in humans

11.1	The gas exchange system	55
11.2	Gas exchange	56
11.3	Breathing	57
11.4	Rate and depth of breathing	58

12 Respiration

12.1	Aerobic respiration	59
12.2	Anaerobic respiration	60

13 Excretion

13.1	Excretion	61
13.2	Kidney structure	62
13.3	Kidney function	63
13.4	Kidney dialysis and kidney transplants	64

14 Coordination and response

14.1	Nervous control in humans	65
14.2	Neurones and reflex arcs	66
14.3	Synapses and drugs	67
14.4	Sense organs	68
14.5	The eye	69
14.6	Hormones	70
14.7	Controlling conditions in the body	71
14.8	Controlling body temperature	72
14.9	Tropic responses	73

Contents

15 Drugs
15.1	Drugs	74
15.2	Heroin	75
15.3	Alcohol and the misuse of drugs in sport	76
15.4	Smoking and health	77

16 Reproduction
16.1	Asexual and sexual reproduction	78
16.2	Flower structure	79
16.3	Pollination	80
16.4	Fertilisation and seed formation	81
16.5	The male reproductive system	82
16.6	The female reproductive system	83
16.7	Fertilisation and implantation	84
16.8	Pregnancy	85
16.9	Antenatal care and birth	86
16.10	Sex hormones	87
16.11	The menstrual cycle	88
16.12	Methods of birth control	89
16.13	Control of fertility	90
16.14	Sexually transmitted infections (STIs)	91

17 Inheritance
17.1	Chromosomes, genes and DNA	92
17.2	Protein synthesis	93
17.3	Mitosis	94
17.4	Meiosis	95
17.5	Inheritance and genes	96
17.6	Monohybrid inheritance	97
17.7	Codominance	98
17.8	Sex linkage	99

18 Variation and selection
18.1	Variation	100
18.2	Mutations	101
18.3	Adaptive features	102
18.4	Natural selection	103
18.5	Selective breeding	104

19 Organisms and their environment
19.1	Energy flow	105
19.2	Pyramids of numbers and biomass	106
19.3	Shortening the food chain	107
19.4	Nutrient cycles	108
19.5	The nitrogen cycle	109
19.6	Populations, communities, and ecosystems	110
19.7	Human populations	111

20 Biotechnology and genetic engineering
20.1	Microorganisms and biotechnology	112
20.2	Enzymes and biotechnology	113
20.3	Fermenters	114
20.4	Genetic engineering	115

21 Human influences on ecosystems
21.1	Food supply	116
21.2	Habitat destruction	117
21.3	Pollution	118
21.4	Water pollution	119
21.5	The greenhouse effect	120
21.6	Acid rain	121
21.7	Sustainable resources	122
21.8	Sewage treatment and recycling	123
21.9	Endangered species	124
21.10	Conservation	125

22 Language focus
126

23 Revision
130

24 Project ideas
134

25 Practical biology
136

26 Mathematics for biology
142

27 Exam-style questions
147

Glossary	156
Answers	158
Data sheets	168

Characteristics and classification of living organisms

1.1 Characteristics of living organisms

Language lab

TRUE or FALSE?

Respiration is a process in which energy is released by the oxidation of foods.

1. The seven characteristics of living organisms are **respiration, growth, sensitivity, nutrition, excretion, movement,** and **reproduction**.

   Complete this table by choosing words from this list and writing them opposite their correct meanings.

	meaning	characteristic
A	the ability to detect stimuli and make appropriate responses	
B	a set of processes that makes more of the same kind of organism	
C	removal from an organism of toxic materials, the waste products of metabolism, or substances in excess of requirements	
D	a set of chemical reactions that breaks down nutrients to release energy in living cells	

   [4]

2. To biologists, classification means

   A   giving organisms a name   B   identifying organisms

   C   putting organisms into groups   D   describing organisms

   Underline your answer. [1]

3. The following is a list of groups that biologists use to classify living organisms

   class   family   genus   kingdom   order   phylum   species

   Rewrite the list in the correct hierarchy of classification.

   [3]

Characteristics and classification of living organisms

1.2 Classification

Language lab

Fill in the gaps.

In the hierarchy of classification, humans belong to the class ....,

the .... *Homo*, and the .... *sapiens*.

1. The drawings show four common birds that came to feed in an English garden.

Parus caeruleus

Parus major

Turdus merula

Erithacus rubecula

   a. State which two birds scientists believe are most closely related and explain your answer.

   ....

   .... [2]

   b. i. Complete this table showing the external features of these birds. One example column has been completed.

Species/feature	All feathers the same colour	Dark stripe along length of body	Large pale areas on sides of head
Erithacus rubecula	X		
Parus caeruleus	X		
Parus major	X		
Turdus merula	✓		

   [3]

   ii. Use the information in this table to complete the following key to identify the four birds.

   1. No large pale areas on head　　　　　　　　　go to 2

      Large pale areas on head　　　　　　　　　　go to 3

   2. All feathers the same colour　　　　　　　　*Turdus merula*

      Feathers of different colours　　　　　　　　*Erithacus rubecula*

   3. ....　　　　..

      ....　　　　..

   [6]

Characteristics and classification of living organisms

1.3 Features of organisms

Language lab

Bacteria fit into the classification group called Prokaryotes. Explain the meaning of this name.

1. The diagram shows a single-celled organism called *Euglena gracilis*. This organism has some features which are usually only found in animals and some which are usually found only in plants.

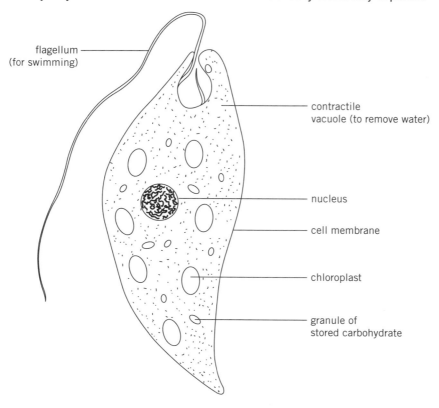

a. i. State **one** feature of *Euglena* which is usually found only in plants.

   .... [1]

   ii. State the characteristic of living organisms which this structure carries out.

   .... [1]

b. i. The contractile vacuole is usually found only in animals. State the characteristic of living organisms carried out by the contractile vacuole.

   .... [1]

   ii. The flagellum is more often found in animals. State the characteristic of living organisms carried out by the flagellum.

   .... [1]

c. All living organisms require a supply of energy. State the name of the process which supplies energy to a cell such as *Euglena*.

   .... [1]

4

Characteristics and classification of living organisms

1.4 Vertebrates

Language lab

Fill in the gaps.

All .... have a backbone. There are five .... in this group, birds, ..., amphibians (none of which have scales), as well as ...., and .... (which do have scales).

1. This table compares some features of **chordate** (**vertebrate**) animals.

   a. Define the term chordate (vertebrate).

   .... [1]

   b. Complete this table.

chordate	body covering	constant body temperature	parental care of young
(fish)		no	no
(frog)	moist skin		no
(turtle)	scales	no	
(bird)	feathers		yes
(rabbit)			yes

[6]

Characteristics and classification of living organisms

1.5 Invertebrates

Language lab

TRUE or FALSE?

1. All invertebrates have jointed legs.
2. Spiders and mites are similar types of insect.
3. All insects have three body parts.
4. Centipedes have at least 100 legs.

1. These four animals were among a group of organisms collected from leaf litter lying on the floor of a deciduous woodland.

Ant   Earthworm   Centipede   Mite

a. Complete the table below to compare the four animals.

	ant	earthworm	centipede	mite
number of pairs of jointed legs present				
are antennae present? (**yes** or **no**)				

[4]

b. Use this key to place each of the animals in its correct group.

1. jointed legs present	go to question 2
no jointed legs	*Annelid*
2. more than four pairs of legs	go to question 3
four pairs of legs or fewer	go to question 4
3. body in two main parts, legs not all alike	*Crustacean*
body made up of many similar segments, with legs alike one another	*Myriapod*
4. 3 pairs of legs present	*Insect*
4 pairs of legs present	*Arachnid*

Write your answers in the table below.

animal	classification group
ant	
earthworm	
centipede	
mite	

[4]

6

Characteristics and classification of living organisms

1.6 Ferns and flowering plants

Language lab

The following scrambled words refer to plants. Unscramble them.

MEST   AVLEEAS   OLESLULEC   RNESF

1. a. Match up the following parts of a plant with the function performed by each of them.

part of plant
stem
root
leaves
flowers
fruit

function
absorb water and mineral ions
usually help dispersal of seed, a reproductive structure
hold leaves in the best position
may be attractive to pollinating insects or birds
trap light energy for photosynthesis

[5]

b. Complete the following paragraphs about the lives of plants. Use words from this list – each word may be used once, more than once, or not at all.

algae   angiosperms   autotrophic   cellulose   chlorophyll
chloroplast   dicotyledons   ferns   herbivorous
monocotyledons   photosynthesis   respiration   starch

All plants contain the light-absorbing pigment called .... This means

that plants can be .... – they can make their own food molecules

from simple inorganic sources by the process of .... All the members

of the Plant Kingdom are made of cells surrounded by a cell wall made of ....

The Plant Kingdom can be divided into four phyla, ..., mosses, ... and

seed plants. Many of the seed plants have the seed enclosed inside a fruit – they are called

...., and exist in two groups .... (which

have leaves with parallel veins) and .... (leaves have branched veins).

[9]

7

Characteristics and classification of living organisms

1.7 Dichotomous keys

Language lab

The most useful information for producing a dichotomous key is (a) an external structural feature and (b) a question which has only two answers (usually YES or NO). Is this statement TRUE or FALSE?

The diagram below shows four species of marine fish.

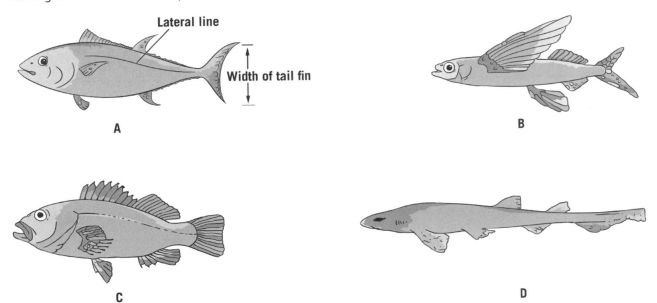

Use this key to identify the four species of fish.

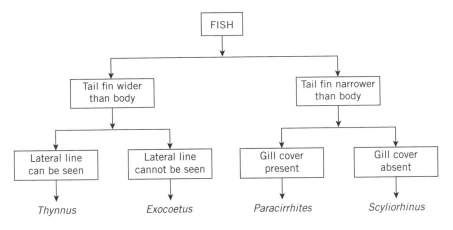

Write your answers in the table:

Letter	Name of species
A	
B	
C	
D	

[3]

Cells 2.1 Structure of cells

Language lab

Explain why plant cells usually have a fixed, regular shape while animal cells are more flexible.

....

....

1. The diagram shows a plant cell and an animal cell.

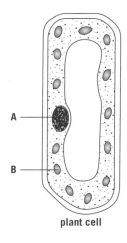
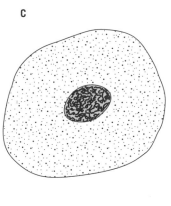

plant cell   animal cell

a. Identify the structures **A** and **B**.

A .... B .... [2]

b. i. Use a guideline to join the letter **C** to a possible position of an organelle which carries out aerobic respiration. [1]

ii. Name this organelle .... [1]

c. The animal cell is actually 25 μm in diameter. Use this information to calculate the magnification of the diagram. Show your working.

[2]

d. State the name of the structure that controls the entry and exit of materials from the plant cell.

.... [1]

Cells   2.2 Cell organelles

> **Language lab**
>
> Match the organelles with their functions.
>
> Organelles: **nucleus**   **ribosome**   **mitochondrion**   **cell membrane**
>
> Functions: **protein synthesis**   **active transport**   **respiration**   **site of DNA**

1. a. The diagrams below show a human nerve cell and a palisade cell from a leaf.

   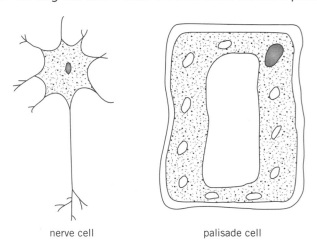

   nerve cell   palisade cell

   i. On the diagrams, label **two** features found in both cells. [2]

   ii. State the name of **one** structure found in the palisade cell which allows it to carry out photosynthesis.

   .... [1]

   iii. State the name of **one** other structure found only in plant cells.

   .... [1]

   b. All living cells are able to release energy by the process of respiration.

   i. State the name of the structures in which aerobic respiration takes place.

   .... [1]

   ii. Energy from respiration may be used in protein synthesis.

   State the name of the structures in which protein synthesis takes place.

   .... [1]

10

Cells

2.3 Different types of cell

Language lab

Match the cells with their functions:

Cells: red blood cell   neurone   phloem sieve tube   root hair cell

Functions: sucrose transport   impulse conduction   ion uptake   oxygen transport

1. The diagrams show several types of plant and animal cells. They are not drawn to the same scale.

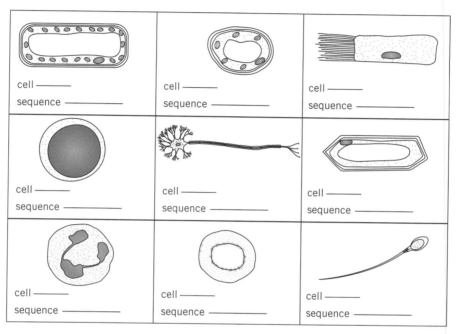

a. Use the key below to identify each of the cells.

Write the letter corresponding to each cell on the line next to the appropriate diagram.

For each of the cells, write down the sequence of numbers from the key that gave you your answer.

Key

1. Cell has clear and obvious cell wall	go to 2	5. Cell has projections at one or more ends	go to 6
Cell has membrane but no cell wall	go to 4	Cell does not have projections	go to 8
2. Cell has chloroplasts in the cytoplasm	go to 3	6. Cell has projections at both ends	**CELL E**
Cell does not have chloroplasts in the cytoplasm	**CELL A**	Cell with projections or projection at only one end	go to 7
3. Cell with fewer than 10 chloroplasts visible	**CELL B**	7. Cell has many projections at one end	**CELL F**
Cell with more than 10 chloroplasts visible	**CELL C**	Cell has a single long projection at one end	**CELL G**
4. Cell contains a nucleus	go to 5	8. Cell has nucleus with many lobes	**CELL H**
Cell does not have a nucleus	**CELL D**	Cell has a round nucleus	**CELL I**

b. Two of the cells that you have identified would be ffound in the same plant organ.

State the name of this plant organ. .... [1]

11

Cells — 2.4 Levels of organisation

Language lab

Which one of **A**, **B**, **C**, or **D** shows the correct order of structures from the simplest to the most complex?

A   organs – systems – tissues

B   systems – organs – tisues

C   tissues – systems – organs

D   tissues – organs – systems

1. Many organisms are made up of cells, tissues, and organs.

   a. The heart, stomach, and bladder are examples of organs found in all mammals.

      State the name of the systems to which each of them belongs.

      Heart ....

      Stomach ....

      Bladder .... [3]

   b. Phloem and stamens are examples of structures found in plants.

      State the name of the systems to which each of them belongs.

      Phloem ....

      Stamens .... [2]

Movement in and out of cells

3.1 Diffusion

Language lab

A student gave the following definition of **diffusion**. Is it correct or incorrect? Explain your answer.

"Diffusion is the movement of molecules, down a concentration gradient, across a partially permeable membrane."

1. Complete these paragraphs about the movement of molecules in and out of cells.
   Use words from this list. Each word may be used once, more than once, or not at all.

cellulose	diffusion	down	gas	liquid	osmosis
partially permeable		potential	random	rapid	through	up

   .... is a process in which molecules move .... a concentration gradient. The movement may take place in a .... or a liquid, and is the result of the .... movement of the molecules. The process continues until an .... is reached.

   .... is the .... of water molecules, and takes place down a water .... gradient. This process occurs across a .... membrane.

   [9]

13

Movement in and out of cells

3.2 Osmosis

Language lab

Fill in the gaps to complete this statement about osmosis.

Osmosis is the .... of .... The process occurs across a .... – plant cells which lose water become .... and plant cells which gain water become ....

1. A group of students carried out an investigation into the effects of sugar solutions on rods of potato. The rods of potato were cut using a cork borer, then gently blotted and weighed.

   The rods were placed in groups of three in dishes of distilled water or in one of several sugar solutions of different concentrations. After 4 hours the rods were removed from the solutions, gently blotted, and reweighed.

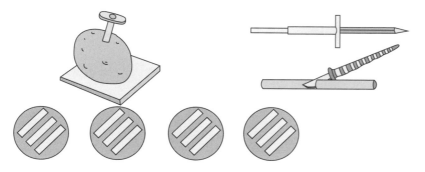

The students converted the raw results into percentage change in mass for each sugar concentration. The results are shown in the table below.

concentration of sugar solution (arbitrary units)	percentage change in mass for rod 1	percentage change in mass for rod 2	percentage change in mass for rod 3	mean percentage change in mass
0 (distilled water)	+8	+8	+8	
0.25	+5	+3	+4	
0.5	0	−2	−1	
1.0	−6	−8	−4	
1.5	−9	−10	−11	

a. i. Calculate the mean percentage change in mass for each of the different solutions. Write your answers in the right-hand column of the table. [1]

   ii. Plot a line graph of the data on the grid to the right.

   iii. From the graph, calculate the concentration of the sugar solution which would result in no change in mass of the potato tuber.

   .... arbitrary units [1]

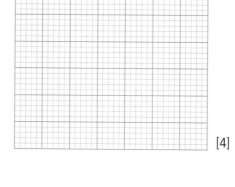

[4]

   iv. Suggest the significance of the value you have obtained for part **iii**.

   .... [1]

b. State the name of the process that causes the change in mass of the potato during the course of the investigation.

   .... [1]

14

Movement in and out of cells

3.3 Osmosis in plant and animal cells

Language lab

Use words from this list to fill in the gaps in the following paragraph.

osmosis　　haemolysis　　turgidity　　flaccid　　cellulose　　starch　　active　　transport

Water is important in plants for photosynthesis and also to prevent the plants becoming .... and wilting. Water enters and leaves plant and animal cells by the process of ....: plant cells can resist the entry of too much water as they have a .... cell wall, but animal cells can suffer .... (they burst as the excess water breaks the fragile cell membrane).

1. Plant tissues can benefit from cells becoming turgid.

   a. Explain the meaning of the term turgid.

   ....

   .... [1]

   b. Explain how turgor can benefit plant tissue. You should use a diagram in your answer.

   [3]

   c. Examine the diagram below. It shows turgid potato tissue placed in a solution.

   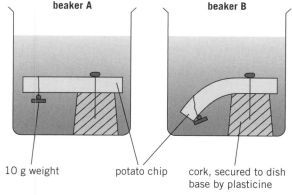

   beaker A　　　　beaker B

   10 g weight　　potato chip　　cork, secured to dish base by plasticine

   i. State which beaker contains pure water. .... [1]

   ii. Describe and explain what would happen if the potato chip from beaker **B** was placed in beaker **A** for 24 hours.

   ....

   ....

   .... [2]

Movement in and out of cells

3.4 Active transport

Language lab

True or false? Underline your answers.

Active transport:

A produces energy

B requires a supply of ions

C requires energy

D moves materials against a concentration gradient

E only occurs in animal cells

1. The diagram shows different ways in which molecules may move in and out of cells. The dots show the concentration of molecules.

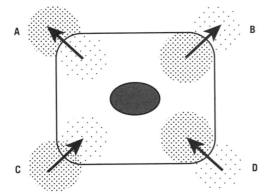

The cell is respiring aerobically.

a. Which arrow represents   i. the movement of carbon dioxide molecules? .... [1]

   ii. the movement of oxygen molecules? .... [1]

b. Which arrow represents the active transport of glucose molecules into the cell? .... [1]

c. Explain your answer to part **b**. ....

   ....

   .... [2]

d. Complete this table to describe **two** other examples of active transport.

example number	name of substance transported	site of transport
1		
2		

[4]

16

Biological molecules

4.1 Biological molecules

Language lab

True or false?

All organic molecules contain carbon, hydrogen, and oxygen.

1. Complete these paragraphs about some biological molecules.

   Use words from this list. Each word may be used once, more than once, or not at all.

   **amino acids   cellulose   common   DDT   DNA   fatty acids   glucose   glycerol
   glycogen   haemoglobin   insoluble   simple sugars   soluble   sucrose**

   a. Starch consists of smaller units called .... Another molecule made up of these subunits is ...., found in plant cell walls. The sugar most often used for sweetening foods is .... which is so useful because it is ....

   b. Fats are made of smaller units called .... linked to one another by .... Fats are often used for energy storage or as barriers between watery environments – a property that is useful is that fats are .... in water.

   c. Proteins such as .... are made of subunits called .... These subunits are .... so that they are easily transported from one part of the body to another.

   d. Nucleic acids such as .... are made of smaller units that act as a code for protein synthesis in cells.

   [4]

2. Complete this table. Place a + if the element is present and a – if it is not.

	carbon	hydrogen	oxygen	calcium	phosphorus	nitrogen
protein						
fat						
carbohydrate						
nucleic acid						

   [4]

Biological molecules

4.2 Chemical tests for biological molecules

Language lab

Match the following terms in pairs.

protein	contain approximately twice as many hydrogen atoms as oxygen atoms
fat	can oxidise DCPIP
carbohydrate	are insoluble in water
vitamin C	enzymes are examples

1. Soya is a vegetarian meat substitute made from soya beans.

   The table below gives information about the food values of 100 g of soya and 100 g of beef.

food	energy content (kj)	fat content (g)	protein content (g)	carbohydrate content (g)
soya	1800	23.5	40.0	12.5
beef	1625	21.0	15.0	6.5

   a. i. Use the information in the table to suggest the main advantage of soya compared with beef.

   .... [1]

   ii. Soya is cheaper to produce than beef. Suggest two reasons why soya is cheaper than beef.

   ....

   .... [2]

   b. The nutrient content of different foods can be investigated with a series of simple chemical tests.

   A group of students was given samples of four different powdered foods. They were also given a sample of pure table salt, which only contains sodium chloride. They carried out three tests, for glucose, starch, and protein. The table shows the colours produced as a result of their tests.

food test	sample **W**	sample **X**	sample **Y**	sample **Z**	table salt
glucose	red/orange	blue	red/orange	blue	blue
starch	brown	blue/black	brown	brown	brown
protein	purple	purple	blue	purple	blue

   i. State which one of the samples contained protein but no starch or glucose. .... [1]

   ii. Suggest why table salt was included in the testing. ....

   .... [1]

   iii. Potato contains starch and protein but not glucose. State which food might have been potato.

   .... [1]

18

Biological molecules — 4.3 DNA

Language lab

The initials DNA stand for .... The structure of this molecule was discovered by two scientists, ... and .... in 1953. They were helped by the X-ray studies of ....

True or false about DNA

1. a. Here is a list of statements about DNA. State whether each of the statements is true or false.

statement	true	false
DNA carries coded instructions for the characteristics of an organism		
DNA is only found in the nucleus of cells of vertebrate animals		
DNA is a large molecule, but its code is carried in only four different subunits		
A DNA profile can be used to identify a criminal		
DNA can be extracted from dead dinosaurs		
Plant cells have different DNA to animal cells		
One difference between animals and bacteria is that bacteria do not have DNA		
The DNA profile is unique to each individual human		
50 human cheek cells fit into a 1 mm space, but each cheek cell contains 2 m of DNA		
Zoo scientists can copy the DNA found in an orang-utan, and check if it is related to an orang-utan at another zoo		

b. Here are three more statements about DNA. Working with two other people (so that there is an odd number in the group), discuss each of these statements. At the end of your discussion, take a vote to decide whether your group agrees or disagrees with the statement. Your teacher might ask you to explain your decision.

statement	agree	disagree
The police should be able to collect DNA from anyone in the country to use for checking crime scenes		
A health insurance company has the right to check your DNA profile before giving you a life insurance policy		
An adopted child should be able to use its DNA profile to track down its biological parents		

Enzymes

5.1 Structure and action of enzymes

Language lab

True or false?

1. All enzymes are proteins.
2. No enzymes can function under very acidic conditions.
3. Enzymes are only found in animal cells.

1. The two lists show some words referring to enzymes, and definitions of these words.

   Draw lines to match up the terms with their definitions.

word
protein
substrate
product
active site
denaturation
optimum

definition
a molecule that reacts in an enzyme-catalysed reaction
the part of the enzyme where substrate molecules can bind
a change in shape of an enzyme so that its active site cannot bind to the substrate
the ideal value of a factor, such as temperature, for an enzyme to work
the type of molecule that makes up an enzyme
the molecule made in an enzyme-catalysed reaction

   [6]

2. This table lists some important uses of enzymes.

name of enzyme	use of enzyme
	part of biological washing powders – removes fatty stains
restriction enzyme	
lactase	
	softens some parts of leather in the clothing industry
	could break down tough plant cell walls
	breaks down starch during germination of seeds

   Complete the table by filling in the gaps, using words or phrases from this list.

   amylase   cellulase   clears pieces of tissue from fruit juices
   cuts out useful genes from chromosomes   lipase   maltase   pectinase
   protease   releases carbon dioxide during respiration   removes milk sugar from milk

   [6]

20

Enzymes

5.2 Factors affecting enzyme action: temperature

Language lab

Complete the definition.

Denaturation is ....
....

1. Rennin is an enzyme which causes milk to clot. It is used in cheese-making to start making the solid curd.

   A student decided to carry out an experiment on the effect of temperature on the clotting of milk by rennin. She wrote down her results on a scrap of paper.

   Temp and clotting of milk
   60 – nothing
   30°C – only 8 mins.
   15 – nothing again (no clot)
   55 – 7 min
   45 – fastest yet! – 2 min.
   35°C – 5 minute
   20 – 35 mins. to clot
   40 – fast – 3 minutes!
   25 – 18 min.
   50 – 5 min. (same as 35°C!)

   a. Present these results in the form of a suitable table. Draw your table in this space or in your note book. [5]

   b. Plot a graph of these results. Use the grid below.

   c. State the optimum temperature for the clotting process ....°C [1]

   d. i. State one other factor which could affect the activity of the enzyme.

      .... [1]

      ii. Suggest how a student could control this other factor.

      .... [1]

21

Enzymes   5.3 Enzymes and pH

Language lab

Fill in the gaps to complete this paragraph about enzymes and pH.

Enzymes have an .... pH, the pH at which the enzyme functions most efficiently. An enzyme which works well at pH 2 could be found in the .... of a mammal, and an enzyme which works well at pH 8 could be found in the .... of a human.

1. A set of experiments was carried out to investigate the effects of pH on two enzymes, pepsin (a protease) and amylase.

   Each reaction was performed at 37 °C and allowed to carry on for 15 minutes. The rate of activity was calculated by measuring the amount of product formed over the time period of the reaction.

   The results are shown in the graph (right).

   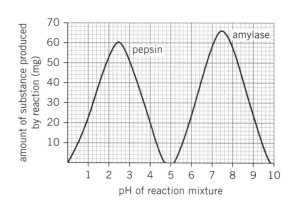

   a. i. Name the products produced by the pepsin-controlled reaction.

      .... [1]

      ii. State how much of this product was formed at pH 2.
      .... mg [1]

   b. State the pH at which 60 mg of substance was produced by:

      i. amylase ....   ii. pepsin .... [1]

   c. Pepsin works best at low pH (acidic conditions). Explain how these conditions are achieved in the human digestive system.

      ....
      .... [1]

   d. Scientists manufacturing Sparklo, a new biological washing powder, were looking for an enzyme which would remove lipids (fats and oils) from clothes.

      i. Name the enzyme that would remove fats and oils from the clothing.

         .... [1]

      ii. Name the products formed as the enzyme digests fats.

         .... [1]

      iii. Explain how a product of the liver increases the digestion of fats.

         Name of liver product .... [1]

         Explanation of effects on fat digestion ....
         ....
         .... [2]

Plant nutrition — 6.1 Photosynthesis

Language lab

Complete this word equation by filling in the blanks.

Carbon dioxide + ... = oxygen + ...

1. Complete the following paragraph about plant nutrition.

   The production of food by plants is called .... This process uses .... energy trapped by .... in the leaves. The process also uses two raw materials from the environment – .... from the air and .... from the soil. The first product that can be easily detected is .... , a storage carbohydrate. Green plants also produce the gas .... during this process – it is released through tiny pores called .... [8]

2. The diagram shows the movement of materials in and out of a leaf during photosynthesis.

   materials in and out of leaf

   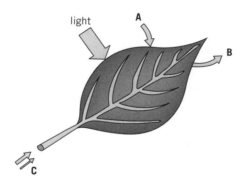

   a. Name the gas entering at **A** .... [1]

   b. Name the gas entering the atmosphere at **B** .... [1]

   c. Name the raw material, required for photosynthesis, entering at **C**.

   .... [1]

   d. Name the mineral, required to manufacture proteins, entering at **C**.

   .... [1]

   e. State the name of the mineral required for the production of the green pigment in the leaves.

   .... [1]

Plant nutrition

6.2 What is needed for photosynthesis?

Language lab

Complete this paragraph.

Photosynthesis produces carbohydrates, including the insoluble .... This molecule can be tested for using .... solution, which changes from .... to .... if the insoluble carbohydrate is present.

1. Some plants have variegated leaves (some parts of the leaf are green and other parts are white). A variegated plant was placed in a dark cupboard for 48 hours. When it was removed it had part of one leaf covered with black card and then it was left exposed to bright light.

   After six hours, small discs were cut from the leaf and each was tested for the presence of starch. The experiment is summarised in the diagram below.

   a. State the reason for placing the plant in a dark cupboard.

   .... [1]

   b. State the name of the mineral required for the production of the green pigment in the leaves.

   .... [1]

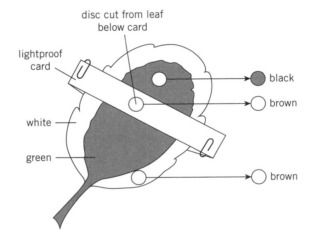

   c. Use the results of the experiment to identify one factor necessary for photosynthesis.

   .... [1]

   Explain why this factor is necessary for photosynthesis.

   .... [1]

Plant nutrition

6.3 Products of photosynthesis

Language lab

True or false?

1. Starch gives a deep orange-red colour with Benedict's solution.
2. Oxygen can relight a glowing wooden splint.

1. a. State the name of the main storage compound in plants.

    .... [1]

    b. Describe how you would carry out a test for this storage compound.

    ....

    ....

    ....

    .... [2]

    c. Describe a positive result for this test. .... [1]

    d. Suggest two uses, other than storage, for the carbohydrate made by photosynthesis.

    ....

    .... [2]

Plant nutrition — 6.4 Rate of photosynthesis

Language lab

Which one of the following factors is least likely to affect the rate of photosynthesis?

A carbon dioxide concentration

B light intensity

C oxygen concentration

D temperature

1. The diagram below shows a method of measuring the rate of photosynthesis of a water plant such as *Elodea*.

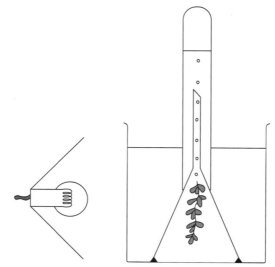

The shoot was exposed to different light intensities and the rate of photosynthesis estimated by counting the number of bubbles released in a fixed amount of time.

The results are shown in this table:

Light intensity (arbitrary units)	1	2	3	4	5	6	7	8
Number of bubbles per minute	6	12	19	25	29	31	31	31

a. Suggest the light intensity at which the plant would have produced 22 bubbles per minute.

.... arbitrary units [1]

b. Explain how the results illustrate the concept of **limiting factors** in photosynthesis.

....

....

.... [2]

Plant nutrition

6.5 Glasshouse production

> **Language lab**
>
> A greenhouse grower may light a paraffin stove in his greenhouse. The most likely reason for this is to:
>
> A remove pest insects
>
> B raise carbon dioxide concentration
>
> C use up excess oxygen
>
> D provide water by condensation

1. A group of scientists working in an experimental plant research station were interested in how different factors play a part in the control of photosynthesis. They made a series of measurements under different conditions. Their results are shown in the table below.

	temperature (°C)	light intensity (arbitrary units)	carbon dioxide concentration (%)	rate of photosynthesis (arbitrary units)
A	22	5	0.04	80
B	22	10	0.04	80
C	22	5	0.15	170
D	22	10	0.15	200
E	30	5	0.04	80
F	30	10	0.04	80
G	30	5	0.15	185
H	30	10	0.15	300

   a. Describe the set of conditions which gave the highest rate of photosynthesis.

   .... [1]

   b. The normal concentration of carbon dioxide in the atmosphere is about 0.04%. Use the results in the table to describe the benefits of adding carbon dioxide to the air.

   ....

   .... [2]

   c. Explain why the rate of photosynthesis is the same in conditions **A**, **B**, **E**, and **F**.

   ....

   .... [2]

Plant nutrition — 6.6 Leaves

Language lab

Gases enter leaves mainly through the

A cuticle   B stomata   C phloem   D epidermis

1. The diagram shows a section through part of a leaf.

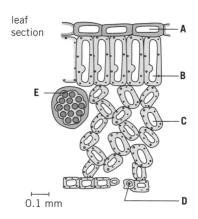

leaf section

0.1 mm

a. Name the cells labelled **A**, **B**, **C**, **D**, and **E**.

letter	cell type
A	
B	
C	
D	
E	

[5]

b. State the letter that identifies the cells where most photosynthesis occurs. .... [1]

c. State the letter that identifies the cells which transport water and minerals into the leaf. .... [1]

d. State the name of the carbohydrate that is transported around the plant.

.... [1]

e. Look back at the leaf section. Use the scale to calculate:

  i. the thickness of the leaf

  ii. the length of a palisade cell.

  In each case, show your working. [4]

28

Plant nutrition
6.7 Mineral requirements

Language lab

Complete this paragraph about mineral ions by filling in the gaps.

Plants require .... to make the green pigment chlorophyll, and .... to build up proteins. Minerals can be absorbed against a concentration gradient by the process of ....

1. The diagrams below show the results of an investigation into the mineral requirements of wheat seedlings.

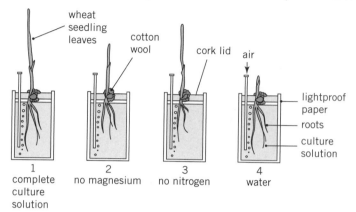

tube number	mineral content	total length of leaves (mm)
1	complete	24.5
2	no magnesium	11.5
3	no nitrogen	20.5
4	no minerals	5.5

a. i. Explain why tubes 1 and 4 were included in the investigation.

....

....

.... [2]

ii. Explain why it is necessary to bubble air through the tubes.

....

....

.... [2]

iii. Explain the results for tube number 2.

....

....

.... [2]

b. Farmers often supply these minerals as fertilisers. The bags of fertiliser often have the letters NPK stamped on them.

State what the initials NPK stand for: .... [1]

29

Animal nutrition

7.1 A balanced diet

Language lab

Most of the energy in our diet comes from:

A vitamin D   **B** protein   **C** carbohydrates   **D** fats

1. a. This pie chart shows the proportion of different food molecules in a diet.

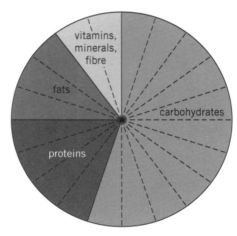

Calculate the percentage of the diet made up of carbohydrates and fats together. Show your working.

[2]

b. Choose whether each of the following statements about food and feeding is true or false.

Statement	True or False
Starch is a carbohydrate found in bread and pasta	
Meat and fish are essential in the diet, as they are the only sources of protein	
Biuret reagent gives a light purple colour with protein	
Oranges and lemons contain vitamin C, and so may help to prevent rickets	
Fats dissolved in alcohol give a milky white colour when mixed with water	
Bacteria can be engineered to provide proteins useful to humans	
Strawberries are a good source of iron	
Vitamin C can be detected using a blue solution called DCPIP	
Assimilation is the process in which foods are transferred from the gut to the blood	
Starch gives an orange colour with iodine solution	
Calcium helps build strong teeth and bones	
Kwashiorkor is a form of malnutrition caused by a low-iron diet	
Too much fat in the diet causes scurvy	
The most common molecule in the human body is water	
Humans cannot eat fungi	

[15]

Animal nutrition

7.2 Sources of nutrients

Language lab

The best source of vitamin C is:

A milk   **B** citrus fruits   **C** bananas   **D** fresh fish

1. a. State the two components of a balanced diet that provide the most energy.

   .... [2]

   b. A young woman was cooking a meal for her friends. She decided to make a pasta dish, using skimmed milk for the sauce. 100 g of the dish contained about 600 kJ.

   State the name of the food class that contributes most energy to this total.

   .... [1]

   Explain your answer ....

   ....

   ....

   ....

   .... [2]

31

Animal nutrition

7.3 Balancing energy needs

> **Language lab**
>
> An imbalance between energy needs and energy intake may cause .... and ....
>
> Fill in the gaps from the following choice of words:
>
> **obesity   hair loss   diabetes   rickets   scurvy**

1. The table below shows the daily energy requirements of different people.

person	energy requirement (kJ per day × 1000)
eight-year-old boy or girl	4
teenage girl	12
teenage boy	14
manual worker (builder)	18.5
patient in hospital	7
male IT worker	11.5
female IT worker	9.5
elderly person (aged 75)	8

a. Plot this information as a bar chart. Use the grid provided.

b. Use data from the table to suggest **two** factors which influence an individual's daily energy requirements.

....

.... [2]

c. Suggest why a female IT worker might have a lower energy requirement than a male IT worker.

.... [1]

Animal nutrition — 7.4 Starvation and nutrient deficiency

> **Language lab**
>
> A deficiency of milk in the diet is most likely to cause ....
>
> Choose your answer from:
>
> **diabetes   scurvy   marasmus   rickets**

1. The table shows the percentage of overweight people in different age groups in a population.

age group	percentage of overweight people	
	male	female
20–24	22	20
25–29	27	22
30–34	35	27
35–39	40	32
40–44	51	39
45–49	65	44
50–54	64	49
55–59	62	52

   a. i. State which sex is more likely to be overweight.

   .... [1]

   ii. State which group has the smallest percentage of overweight people.

   .... [2]

   b. i. State the form in which large amounts of extra weight are stored in the human body.

   .... [1]

   ii. Suggest **three** medical conditions associated with being overweight.

   .... [3]

   c. Being overweight is one example of malnutrition. Define the term *malnutrition*.

   .... [2]

   d. Children in less economically developed countries may suffer from marasmus or kwashiorkor. These conditions are together known as **Protein Energy Malnutrition (PEM)**.

   i. Suggest why a shortage of **protein** can be harmful. ....

   .... [2]

   ii. Suggest the likely effects of a shortage of **energy** foods in the diet.

   ....

   .... [2]

33

Animal nutrition — 7.5 Digestion

Language lab

Fats are digested to fatty acids and glycerol. Which one of the following speeds up this change?

A amylase   B insulin   C lipase   D protease

1. Human salivary amylase can break down starch. In order to investigate the optimum conditions for starch digestion, four test tubes were set up as shown in the diagram below.

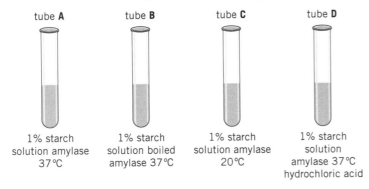

tube **A** — 1% starch solution amylase 37 °C

tube **B** — 1% starch solution boiled amylase 37 °C

tube **C** — 1% starch solution amylase 20 °C

tube **D** — 1% starch solution amylase 37 °C hydrochloric acid

a. After ten minutes, samples were taken from each of the tubes and tested separately with iodine solution and with Benedict's reagent.

Complete the table below to show the likely results of this investigation.

	colour in tube **A**	colour in tube **B**	colour in tube **C**	colour in tube **D**
tested with iodine solution				
tested with Benedict's reagent				

[4]

b. State where amylase is secreted in the human digestive system.

.... [2]

c. Explain why tubes **A**, **B**, and **C** were incubated at 37 °C.

.... [1]

d. Suggest where the conditions in tube **D** are likely to occur in the human digestive system.

.... [1]

Animal nutrition — 7.6 Teeth

Language lab

Match the following pairs.

molar		biting
incisor		killing/piercing
canine		crushing

1. This diagram shows a section through a molar tooth.

   a. i. Name the parts **A**, **B**, and **C**.

   A ....
   B ....
   C .... [3]

   ii. Name **two** structures found in the pulp cavity.

   1 ....

   2 .... [2]

   b. The diagrams below show three stages in tooth decay.

   stage 1 — enamel attacked, no pain
   stage 2 — decay develops, no pain
   stage 3 — decay develops further *PAIN*

   i. Suggest why it may take several **months** for a cavity to develop in the enamel of a tooth.

   .... [1]

   ii. Suggest why pain is not felt until stage 3.

   ....

   ....

   .... [2]

   c. Fibres hold the tooth into its socket in the gum. These fibres loosen and teeth may fall out as a symptom of the disease scurvy.

   State the name of the vitamin needed to produce these fibres properly.

   .... [1]

35

Animal nutrition — 7.7 Mouth, oesophagus, and stomach

Language lab

Fill in the gaps.

In the mouth .... is secreted. This liquid contains the enzyme ....
and the solute .... which makes the pH more ....

1. The diagram shows some stages of ingestion of food.

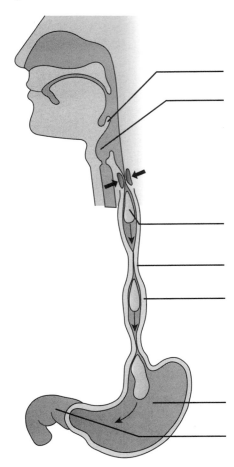

a. Add labels to the diagram, using these words and phrases:

   **bolus of food   circular muscle   contracted circular muscle   epiglottis
   entrance to trachea   relaxed circular muscle   pyloric sphincter   stomach** [7]

b. List **three** functions of saliva.

   1. ....
   2. ....
   3. .... [3]

Animal nutrition

7.8 Small intestine and absorption

Language lab

The surface area of the small intestine is increased by the presence of:

A cilia B flagella C root hairs D villi

---

1. Study the structure shown in the diagram (right).

   a. Name this structure. .... [1]

   b. Suggest where this structure would be found.

      .... [1]

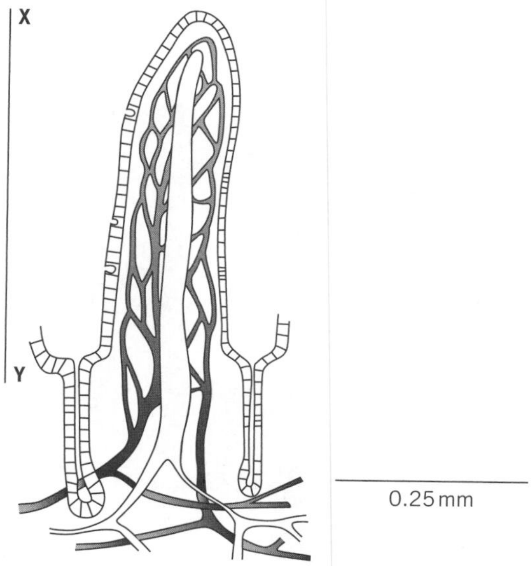

   c. i. Draw a line, labelled M, to show the vessel into which fatty acids and glycerol are absorbed. [1]

      ii. Draw a line, labelled N, to show a cell which releases a product to protect the lining of this structure. [1]

   d. Describe how this structure is adapted to its function.

      ....

      ....

      ....

      .... [3]

   e. Use the scale line on the diagram to calculate:

      i. the magnification of this diagram

      [2]

      ii. the actual height of the structure, from **X** to **Y**.

      Show your working in each case. [2]

37

Animal nutrition — **7.9 Large intestine and intestinal disease**

> **Language lab**
>
> Complete the following sentence.
>
> The large intestine can be divided – in the correct order – into the ...., the ...., the ...., and the ....

1. The diagram below shows the large intestine of a human.

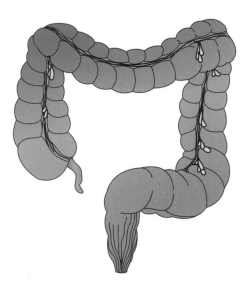

   a. Use guidelines and labels to identify:

      i. a region in which water is reabsorbed

      ii. a region in which faeces are stored

      iii. the appendix. [6]

   b. The large intestine may become infected by a bacterium.

      i. Name one disease of the large intestine caused by a bacterium.

      .... [1]

      ii. Describe and explain one symptom of this disease.

      ....
      ....
      .... [2]

   c. Explain what is meant by oral rehydration therapy (ORT).

      ....
      ....
      .... [2]

Plant transport

8.1 Transport systems

> **Language lab**
>
> The following words refer to plant transport, but they are scrambled up. Unscramble the words.
>
> LHPEMO   PIEREMISD   MXYLE   CRUOSSE
>
> When you have unscrambled the words, match them with their definitions.
>
> A main carbohydrate transported in plants
>
> B site of water and mineral transport
>
> C outer covering of plant
>
> D tissue which transports sugars

1. Study the diagrams below.

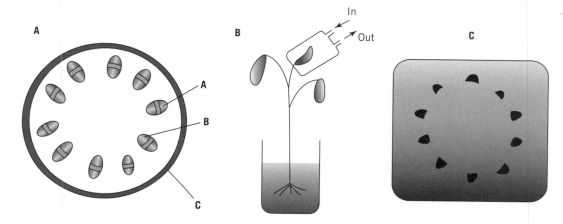

   a. Identify the tissues labelled **A**, **B**, and **C** in diagram **A**.

   A ....   B ....   C ....   [3]

   b. An experiment was carried out using the plant shown in diagram **B**. The roots of the plant were allowed to stand in a solution of eosin (a red dye), and one of the leaves was kept inside a container that could be filled with radioactively labelled carbon dioxide.

   The apparatus was placed in bright light for 6 hours. A cross-section of the stem was then cut using a sharp scalpel.

   i. State which of the tissues **A**, **B**, or **C** would be stained red. .... [1]

   Explain your answer. ....

   .... [1]

   ii. The section was allowed to stand on a piece of film sensitive to radiation.

   When the film was developed it appeared as shown in diagram **C**.

   Explain why the film had this appearance. ....

   ....

   .... [2]

Plant transport

8.2 Water uptake

Language lab

Fill in the gaps.

Plants need water for .... and as a raw material for .... A plant takes up water through its .... and loses it through the ....

1. These paragraphs refer to transport systems in plants.

   Use words from this list to complete the paragraphs. You may use each word once, more than once, or not at all.

active transport	cambium	diffusion	digestion	epidermis	hairs	ions
magnesium	nitrate	osmosis	phloem	photosynthesis	respiration
solvent	support	surface area	transpiration	vascular	xylem

   Water is obtained by plants from the soil solution. The water enters by the process of .... through special structures on the outside of the root called root .... These structures increase the .... of the root. As well as absorbing water they can also take up .... such as .... which is required for the production of chlorophyll. These substances are absorbed both by .... and by .... (a process that requires the supply of energy).

   Plant cells rely on water for ...., as a .... in chemical reactions, and as a raw material for .... Water is also used as a transport medium. [10]

Plant transport 8.3 Transpiration

Language lab

True or false?

Transpiration causes loss of water and minerals from a plant. ✓

1. The diagram shows the apparatus used by a student in a laboratory investigation of the water balance of a green plant. The roots were carefully washed before the plant was placed in the measuring cylinder. The apparatus containing the plant was weighed at the start of the investigation and again 24 hours later. The scale on the measuring cylinder was used to read the volume of water. The same investigation was also carried out by four other students.

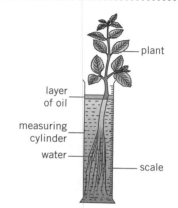

The results of the investigation are shown in the table.

	mass of apparatus / g					mean mass / g	volume of water / cm³					mean volume / cm³
start	220	225	217	221	219		100	100	100	100	100	
24 h later	208	214	209	210	210		87	89	88	87	89	

a. Calculate the mean loss of mass due to water loss from the plant during the 24-hour period. Show your working.

.... [3]

b. Calculate the mean **mass** of water, which has been absorbed by the roots of the plant during the 24-hour period. Show your working.

........................g [3]

c. Using your knowledge of how water moves upwards in a green plant, explain why your answers to **a.** and **b.** are quite similar.

....

....

....[2]

d. Explain why the mass of water absorbed by the roots is not exactly the same as the amount of water lost by the plant.

....

....[2]

41

Plant transport

8.4 Translocation

Language lab

The contents of the translocation system can be sampled by insects, called ...., which feed on the transported sugars.

1. a. The diagrams show transverse sections of a stem (**A**) and a vascular bundle in a leaf (**B**).

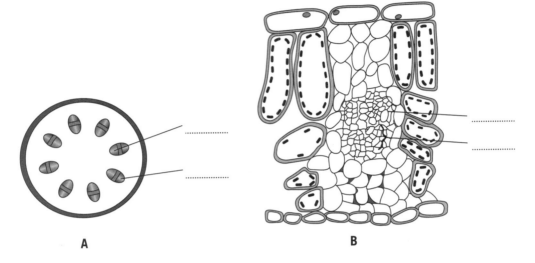

A

B

On each of the diagrams label the phloem and the xylem. [4]

b. The diagram below shows a plant in the sunlight. Three areas of translocation are labelled.

i. Add arrowheads to each of the three 'translocation' lines to show the direction of translocation in the three areas. [3]

ii. Name the main material which is translocated through the body of the plant.

.... [1]

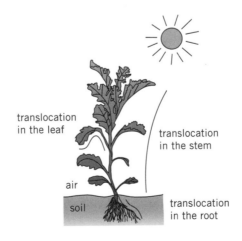

Transport in humans — 9.1 Circulation

Language lab

Match the following pairs.

aorta	from heart to lungs
renal artery	blood input to kidney
hepatic vein	blood output from liver
pulmonary artery	main artery in the body

1. The diagram shows a plan of the mammalian blood system.

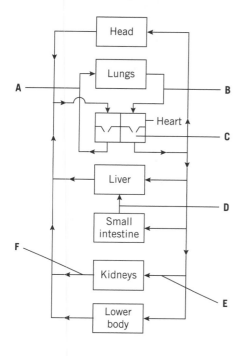

a. i. the name of blood vessel **B** is .... [1]

   ii. Suggest **two** ways in which the blood in vessel **A** differs from the blood in vessel **B**.

   1. ....

   2. .... [2]

b. Suggest **two** ways in which the blood in vessel **E** differs from the blood in vessel **F**.

   1. ....

   2. .... [2]

c. On the diagram, use a guide line and the letter **P** to show the region where the blood is at its highest pressure. [1]

Transport in humans 9.2 The heart

Language lab

True or false?
1. The semilunar valves are at the exits from the ventricles.
2. The left ventricle has thinner walls than the right ventricle.

1. The diagram shows a human heart that has been cut across in the region of the ventricles.

   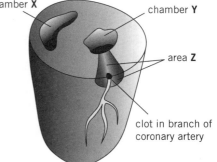

   a. i. State the name of the chamber labelled **Y**. .... [1]

      ii. Explain your answer.

         ....

         .... [2]

   b. The beating of the heart is controlled by a patch of tissue called a ....

      In a healthy person the heart normally beats at about .... beats per minute.

2. a. In mammals, the blood flows through the heart twice for each complete circuit of the body – this is called a

      .... circulation. [1]

   In contrast, in fish blood flows through the heart only once for each complete circuit.

   The diagram shows the difference between these two types of circulation.

   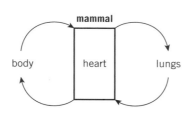 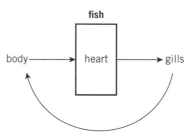

   b. State one **disadvantage** of the single circulation.

      .... [1]

   c. One **advantage** of the double circulation is that pressure is high enough to allow efficient filtration of the blood

      in the paired .... [1]

   d. The veins of fish tend to be much wider than those of mammals.

      Suggest one reason for this. .... [1]

   e. In a mammal, blood is transported to the lungs in the ....

      .... If pressure in these vessels is too high, ...
      can leak into the lungs. This sometimes happens to climbers at high altitude so that the climbers have difficulty
      in breathing. [2]

44

Transport in humans — 9.3 Heart and exercise

> **Language lab**
>
> During exercise the heart pumps more blood to the body. To do this there are two changes to the heart's action. Suggest what these two changes are.

1. The graph shows the pressure changes in the left side of the human heart, during one complete beat.

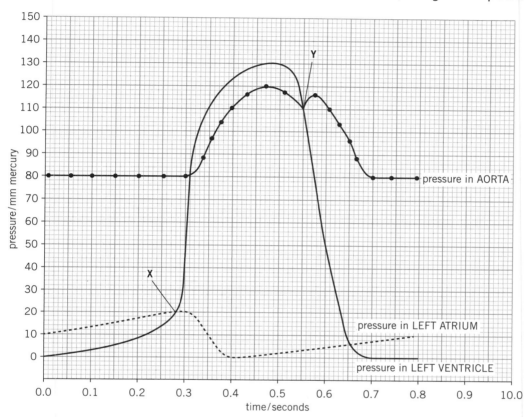

   a. Apart from a change in heart rate, how else can an athlete increase the amount of blood pumped per minute?

   .... [1]

   b. i. From your knowledge of how the heart works, suggest which valves close at points **X** and **Y**. In each case, explain why the valve closes at that point.

   ....

   .... [4]

   ii. Where else in the circulatory system are valves found?

   .... [1]

   iii. What is the purpose of these other valves?

   .... [1]

   c. Suggest why, if the heart is working poorly, a person may have blue lips.

   .... [2]

Transport in humans — 9.4 Blood vessels

Language lab: A doctor usually takes a blood sample for testing from one of a patient's veins. Suggest why the doctor does not take the sample from an artery.

1. The diagram shows changes in blood pressure, speed of blood flow, and the total cross-sectional area of different blood vessels.

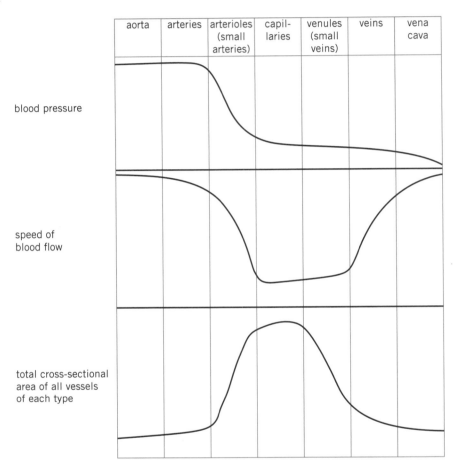

   a. i. Name the blood vessel which carries blood at its highest pressure out of the heart.

   .... [1]

   ii. State in which type of blood vessel the blood loses most of its pressure.

   .... [1]

   b. i. Name the type of blood vessel where materials are exchanged between the blood and the tissue fluid surrounding the body cells.

   .... [1]

   ii. Use information in the diagram to explain two features of this type of blood vessel which help in the exchange of materials.

   1. ....

   2. .... [2]

Transport in humans

9.5 Coronary heart disease (CHD)

Language lab

Which of the following does not cause problems with the heart and circulation?

A high blood cholesterol

B high blood salt concentration

C low vitamin D concentration

D high protein in the diet

1. a. The table shows the rate of blood flow (cm³ per minute) to organs of the human body, at rest and during exercise.

organ	rate of blood flow (cm³ per minute)	
	rest	exercise
heart	300	400
muscles	1000	4500
gut	1500	1000
kidneys	1100	900
skin	450	1500
brain	800	800
total output	5150	9100

   i. State the effect of exercise on the flow of the blood to the organs of the body.

   ....

   .... [3]

   ii. Explain why the change in blood flow to the muscles is important.

   ....

   .... [2]

b. Regular exercise reduces the risk of heart disease.

   i. Explain why a blockage in the coronary artery can lead to a heart attack.

   ....

   .... [2]

   ii. State **two** other factors which increase the risk of heart disease.

   ....

   .... [2]

47

Transport in humans — 9.6 Blood

Language lab

True or false?

1. We have more white blood cells than red blood cells.
2. We need iron to make our blood.
3. We replace our red blood cells every four months.

1. **a.** The diagram below shows the composition of human blood.

   i. Calculate the percentage of plasma in the blood.

   .... [1]

   ii. Name two substances transported in the plasma.

   .... [2]

   iii. Match each of these blood cells with its correct function.

cell type		function
red blood cell		engulfing invading microbes
phagocyte		transport of oxygen
lymphocyte		part of clotting process
platelet (cell fragments)		antibody production

   [4]

   b. The number of red blood cells tends to increase with altitude.

   Suggest the symptoms that might be experienced by a climber being flown by helicopter directly, without training, to 3000 m.

   ....

   .... [2]

   c. Athletes who compete in races over distances from 1500 to 10 000 m do better if they live or train at altitudes greater than 2500 m.

   Suggest a reason for their improved performance.

   .... [1]

   d. Some racing cyclists have used the drug EPO to increase the number of red blood cells they have.

   State the name of one other drug which has been used to 'cheat' in sport. Explain the biological reason for an athlete using this drug.

   Drug ....

   Reason for use .... [2]

Transport in humans

9.7 Blood in defence

Language lab

Which of the following is not a function of white blood cells?

A production of antibodies C engulfing bacteria

B clotting of blood D recognising viruses

1. Match the following words to their definitions.

word
pathogen
transmissible disease
antigen
skin and nasal hairs
active immunity
passive immunity
phagocyte
lymphocyte

definition
external barriers to infection
defence against a pathogen by antibody production in the body
white blood cell which engulfs pathogens
short term defence by provision of antibodies from another individual
a disease-causing organism
a substance which triggers the immune response
cell which produces antibodies
condition in which the pathogen can be passed from one host to another

[8]

2. This graph shows the level of antibody in the blood in response to two doses of a vaccine.

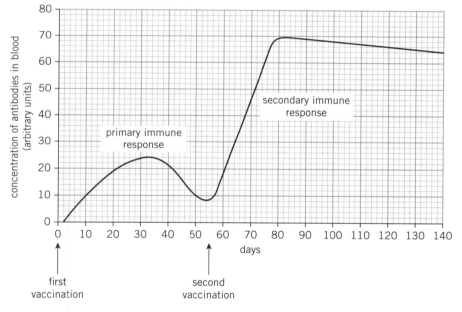

a. Define the term **vaccine**. ....

....

.... [2]

b. State **two** ways in which the response to the second injection is different to the response to the first injection.

....

.... [2]

49

Transport in humans   **9.8 Lymph and tissue fluid**

> **Language lab**
>
> True or false?
>
> 1. Capillaries are the smallest blood vessels.
> 2. Blood pressure rises from entering to leaving the capillaries.
> 3. Water leaves the capillaries by active transport.

1. a. Blood circulation in a mammal is described as a **double circulation**.
   Explain the meaning of the term double circulation.

   ....

   .... [2]

   b. The diagram shows one part of the circulatory system. The blood pressure, in kPa, is marked at three points on the diagram.

   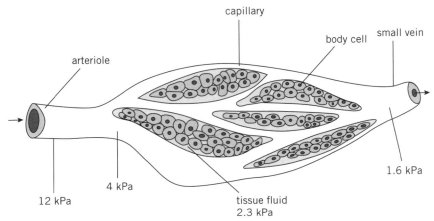

   i. Describe what happens to the blood pressure as blood passes through the arteriole.

   ....

   .... [2]

   ii. State the names of two substances which pass from blood plasma to tissue fluid, and two substances which pass from tissue fluid to blood plasma.

   From blood plasma to tissue fluid .... [2]

   From tissue fluid to blood plasma .... [2]

   c. Suggest what would happen to the blood pressure if a human had a deep cut in the arm.

   .... [1]

   d. Suggest one effect of having a low blood pressure.

   .... [1]

   e. List two features of a capillary that lead to efficient transfer of solutes between blood and tissue fluid.

   ....

   .... [2]

Diseases and immunity

10.1 Disease

> **Language lab**
>
> An organism which causes disease is:
>
> A a bacterium
>
> B a fungus
>
> C a predator
>
> D a pathogen

1 a. State **one** symptom of a person suffering from food poisoning.

   .... [1]

   b. Describe how a mother could spread food poisoning bacteria to her child.

   ....

   ....

   .... [3]

   c. Suggest **one** way a mother could make sure that she did not spread bacteria to her child.

   .... [1]

2. Not all diseases are spread from person to person. These diseases are **non-transmissible diseases**.

   Non-transmissible diseases may be caused by a number of factors.

   Suggest **one** disease which may be caused by **each** of the following factors.

   Deficiency in diet .... [1]

   Inheritance .... [1]

   Age-related degeneration .... [1]

   Lifestyle .... [1]

51

Diseases and immunity   10.2 Defence against disease

Language lab

Complete this sentence.

A second infection by a pathogen brings about a faster response because of the presence of .... cells.

1. The diagram below shows an action taken by one group of blood cells against infection by a pathogen.

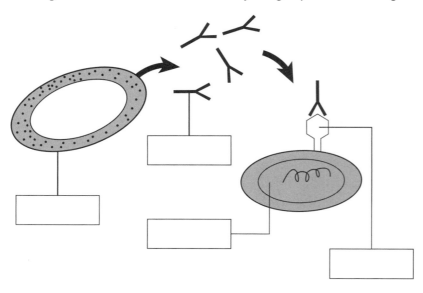

   a. Complete the diagram using words from the following list:

   antibody   antigen   antibiotic   leucocyte   lymphocyte   pathogen

   [4]

   b. Suggest **two** ways in which the invading pathogen may be destroyed.

   1. ....

   ....

   2. ....

   .... [2]

   c. Doctors can protect the body by providing molecules which stimulate the immune response.

   Name this process of protection .... [1]

Diseases and immunity 10.3 Aspects of immunity

> **Language lab**
>
> Immunity can safely result from injection of:
>
> A a pathogen
>
> B a vaccine
>
> C an antibiotic
>
> D blood serum

1. A vaccine has been developed against the organism that causes TB (tuberculosis). The vaccine has been available for use in many parts of Africa from late in the 20th century.

    a. Vaccination against TB is an example of **active immunity**.

    Define the term active immunity. ....

    ....

    .... [2]

    b. State **three** ways in which **passive immunity** differs from active immunity.

    ....

    ....

    .... [3]

    Complete the table below to show whether each method is active or passive.

method of gaining immunity	type of immunity (active or passive)
vaccination against meningitis	
antibodies cross the placenta and enter the fetus	
a person bitten by a dog is given a serum	
a baby feeds on colostrum (first milk)	
a person becomes infected with a pathogen	

Diseases and immunity — 10.4 Controlling the spread of disease

> **Language lab**
>
> A bacterial infection can be controlled with:
>
> A An antibiotic
>
> B An antiseptic
>
> C A disinfectant
>
> D An antidote

1. The pie chart shows the number of food poisoning incidents reported from different food sources.

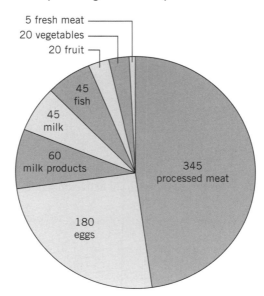

   a. Complete this table using the information in the pie chart.

food source	number of reported outbreaks
eggs	180
	45
fruit	
milk	
	60
processed meat	345
	20

   [3]

   b. Processed meat products include chicken, ham, and corned beef. Packaging on these products usually includes the instruction 'keep refrigerated'.

   Explain why refrigeration makes processed food less likely to cause food poisoning.

   ....

   .... [2]

Gas exchange in humans — 11.1 The gas exchange system

Language lab

Fill in the gaps.

As air leaves the lungs during exhalation, it passes through the ....,

the ...., and the .... before entering the mouth and moving

into the atmosphere.

1. The diagram shows a section through the thorax (chest).

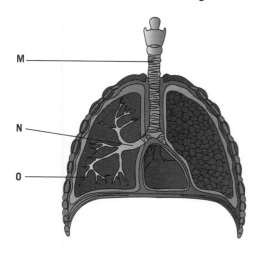

a. Name the structures labelled **M**, **N**, and **O**.

M ....

N ....

O .... [3]

b. On the diagram, use labels and guidelines to identify two different muscles. [4]

c. When muscles contract, they use energy supplied by respiration.

Give **two** other different ways in which this energy may be used.

1. ....

2. .... [2]

d. Exhaled air differs from inhaled air. Name two gases which are present in higher concentration in exhaled air.

.... [2]

Gas exchange in humans   11.2 Gas exchange

Language lab

Fill in the gaps.

Oxygen enters the blood at the .... – it crosses the walls by the process of

.... At the same time, the gas .... leaves the blood.

1. This diagram shows an alveolus (airsac) and a blood capillary.

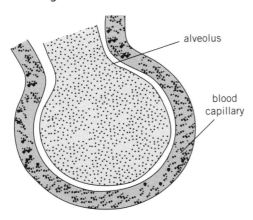

a. On the diagram, use arrows and suitable labels to describe the movement of oxygen and carbon dioxide during gas exchange. [2]

b. List **three** features which make alveoli adapted to gas exchange.

   1. ....

   2. ....

   3. .... [3]

c. State the name of the artery that brings deoxygenated blood to the capillary.

   .... [1]

Gas exchange in humans — 11.3 Breathing

Language lab

True or false?

1. Exhalation uses more energy than inhalation.
2. Breathing releases energy from food.

---

1. The diagram shows the position and shape of the human thorax (chest) during inhalation (breathing in).

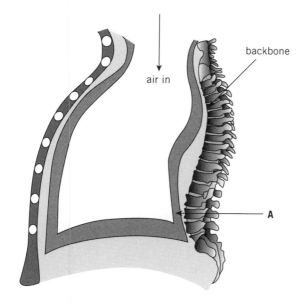

   a. State the name of the part labelled **A**.

   .... [1]

   b. On the diagram label the diaphragm. Use a guideline and the letter **D**. [1]

   c. The diaphragm contains muscles involved in breathing. Name another set of muscles involved in breathing in.

   .... [1]

   d. Describe and explain the how muscles are involved in exhalation (breathing out).

   ....

   ....

   ....

   ....

   .... [3]

Gas exchange in humans — 11.4 Rate and depth of breathing

Language lab

A young athlete took 20 breaths per minute - the volume of each breath was 2000 cm³. At each breath 20% of the inspired air was oxygen and 16% of the expired air was oxygen. The volume of oxygen absorbed per minute was:

A 40 cm³

B 1600 cm³

C 4000 cm³

D 16000 cm³

1. A student breathed into a special 'breathing bag' five times. The bag was empty at the start of the measurement.

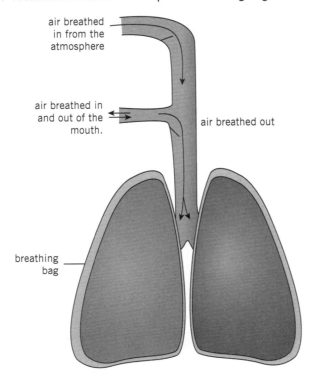

   a. After breathing out five times the volume of air in the bag was measured. The volume was 2500 cm³.

   The student then carried out some exercise for three minutes, and the volume in the bag was measured. This time there was 8000 cm³ of air.

   Use this information to describe the effect of exercise on breathing. ....

   ....

   .... [2]

   b. The rate of breathing is controlled by the concentration of a gas which increases during exercise.

      i. Name this gas .... [1]

      ii. Name the process which removes this gas from the blood .... [1]

Respiration

12.1 Aerobic respiration

Language lab

An athlete may use more than 0.5 kg ATP per minute. Which one of the following statements is true of ATP?

A It is used to remove carbon dioxide
B It is used to pick up oxygen
C It is broken down to release energy
D It is a vital food supply

1. This piece of apparatus is used to study gas exchange.

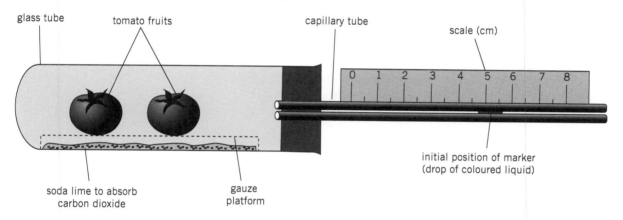

a. During the course of the experiment the marker moves.

   i. State the name of the process, carried out by the cells of the tomatoes, which causes this movement.

   .... [1]

   ii. State the direction in which the marker will move. .... [1]

   iii. Explain your answer to part **ii**. ....

   .... [1]

   iv. Calculate the position of the centre of the marker after 40 minutes, if the marker moves 0.25 cm every five minutes. Show your working.

   [2]

   v. Suggest a suitable control for this experiment.

   ....

   .... [1]

b. The class teacher told the students that fruit growers often enclosed the fruits they collect in plastic bags filled with nitrogen gas.

   Explain why this is important.

   ....

   ....

   .... [2]

59

Respiration

12.2 Anaerobic respiration

Language lab

A product of anaerobic respiration in muscles is:

A water

B carbon dioxide

C lactic acid

D alcohol

1. During exercise muscle contraction requires energy. Some of this energy is released during aerobic respiration.

   a. i. Write a chemical equation for aerobic respiration.

   [3]

   ii. Suggest why respiration is affected by temperature.

   .... [1]

   b. If the muscles work very hard the cells may use up all of the available oxygen. If this happens, energy can still be released by anaerobic respiration. These pie charts show the relative amounts of aerobic and anaerobic respiration during a 100 m race and during a 1500 m race.

   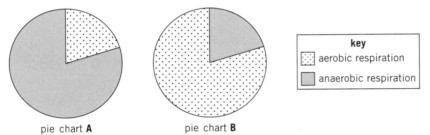

   i. State and explain which of the pie charts shows the results from a 1500 m race.

   ....
   ....
   .... [2]

   ii. State the effect of a build-up of lactic acid on an athlete's performance.

   ....
   .... [2]

   iii. Explain how the level of lactic acid in the blood is reduced after a race.

   ....
   ....
   .... [2]

   c. Explain what is meant by an **oxygen debt**.

   ....
   .... [2]

Excretion

13.1 Excretion

Language lab

Removal of the waste products of metabolism is called:

A egestion

B deamination

C defecation

D excretion

1. a. Define the term **excretion**. ....

.... [2]

b. Complete this table about human excretory products.

name of excretory product	site of production	process responsible for product	site of removal of product
carbon dioxide			
		deamination	
sodium chloride		excess in diet	

[3]

c. The diagram below shows some of the organs in the abdomen, and the blood vessels which supply them.

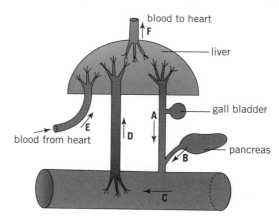

i. State one important function of the liver. ....

.... [1]

ii. Look at the table below. Use guidelines to link each letter with the name of the appropriate structure and a statement about its contents.

letter
A
B
C
D
E
F

structure
hepatic artery
small intestine
bile duct
hepatic portal vein
pancreatic duct
hepatic vein

contents
food
enzymes including amylase and protease
dissolved foods
bile
oxygenated blood
deoxygenated blood

61

Excretion
13.2 Kidney structure

Language lab

Which one of the following systems includes the kidneys?

A circulatory

B digestive

C reproductive

D excretory

1. The diagram shows part of the structure of a kidney, and a single cell from region **A**.

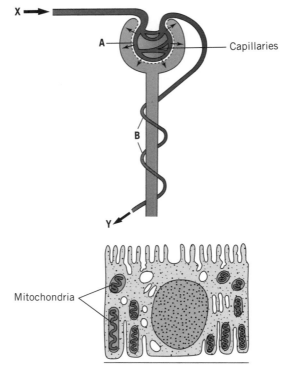

a. Name the vessel supplying blood at **X**. .... [1]

b. Name the ball of capillaries shown in the diagram. .... [1]

c. Explain why glucose can cross the wall at **A** but red blood cells cannot.

   .... [1]

d. Suggest **two** features of the single cell from region **A** which are adaptations to its function.

   1. ....

   2. .... [2]

e. The action of the kidney is to produce urine. Name the tube which carries urine away from the kidney.

   .... [1]

62

Excretion

13.3 Kidney function

Language lab

True or false?

Urea is produced in the liver and removed via the kidneys.

1. Three test tubes were set up as shown in the diagram below.

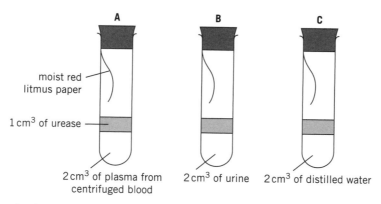

The enzyme urease breaks down urea to release ammonia.

$$urea \longrightarrow carbon\ dioxide + ammonia$$

The tubes were incubated at 37 °C for 30 minutes. After 30 minutes the contents of the tubes are subjected to several tests. The results are shown below.

test substance	A	B	C
biuret reagent	blue to violet	remains blue	remains blue
Benedict's reagent	blue to orange-red on heating	no change on heating	no change on heating
moist red litmus paper	turns blue	turns blue	no change

a. State the name of the substance which the tests show is present in the plasma and in the urine.

.... [1]

b. Explain why the test tubes were incubated at 37 °C.

....

.... [2]

c. i. Identify which substances are found in blood plasma but not in urine or distilled water.

.... [2]

ii. Explain why these substances are not lost from the body.

....

.... [2]

d. State the function of tube **C**.

.... [1]

63

Excretion
13.4 Kidney dialysis and kidney transplants

Language lab

A patient with a transplanted kidney may need immunosuppressive drugs. Why are these drugs needed?

1. Dialysis can help people whose kidneys do not function efficiently. Their blood goes through an artificial kidney machine, where it passes alongside a series of partially permeable membranes. On the other side of the membranes is a solution into which waste substances diffuse.

    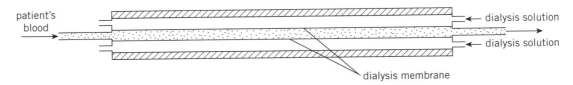

    a. The tube of dialysis membrane is usually coiled rather than straight. Explain why this is important.
    .... [1]

    b. The tube leading out of the machine is usually narrower than the tube leading from the patient's blood into the machine.

       State what effect this will have. ....
       .... [1]

    c. This table compares the concentration of various substances in the blood and the urine of a healthy person.

	plasma concentration (g per 100 cm³)	urine concentration (g per 100 cm³)	
water	910	950	
protein	75	0	
glucose	1.0	0	
urea	0.3	20.5	
sodium ions	3.1	3.7	
chloride ions	3.5	6	

       i. State how the composition of the plasma of a patient requiring dialysis would differ from that of a healthy person.
          .... [2]

       ii. Which of the following substances would be able to cross the dialysis membrane? Underline your choice(s).

           ammonia   glucose   protein   red blood cells   urea
           [1]

       iii. Suggest the composition of a suitable dialysis fluid. Write your answers in the shaded column of the table above. You need only to write **higher, lower**, or the **same**. [6]

64

Coordination and response

14.1 Nervous control in humans

Language lab

Fill in the gaps.

The material which covers an axon is called ...., and it acts as an electrical .... during the conduction of an ....

1. This diagram shows a single neurone.

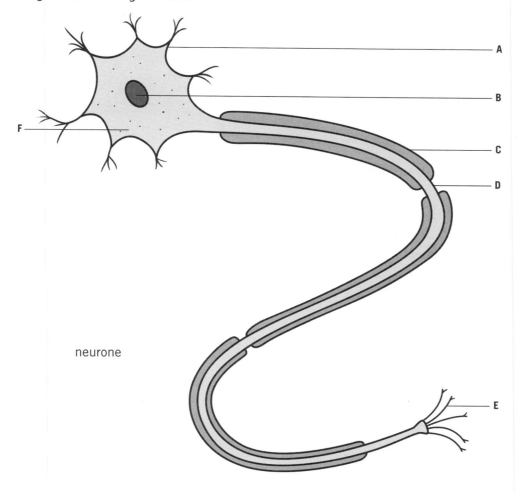

neurone

Match the letters to the descriptions of each of these parts.

description of part	letter corresponding to part
connects with another neurone	
insulates the neurone to prevent interference with other neurones	
contains high concentrations of neurotransmitters	
connects with an effector, such as a muscle	
allows an impulse to 'jump' quickly along the axon	
contains DNA	

[5]

Coordination and response

14.2 Neurones and reflex arcs

Language lab

Coughing, sneezing, and blinking are all examples of reflexes. Suggest **two** features which all reflexes show.

1. a. i. Sneezing is a reflex action. Name the organ containing the receptor cells which detect the stimulus that leads to sneezing.

   [1]

   ii. Suggest **one** advantage of reflex actions.

   ....

   .... [1]

   b. i. The diagram below shows a reflex arc.

   Select words from this list to name the parts numbered 1–7.

   association neurone   dorsal root   grey matter   motor neurone   receptor
   sensory neurone   spinal cord   synapse   ventral root   white matter

structure	name of structure
1	
2	
3	
4	
5	
6	
7	

   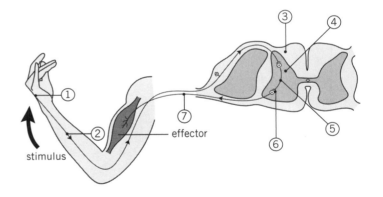

   [7]

   ii. Muscles and glands are effector organs. State how muscles and glands react when they are stimulated.

   Muscles ....

   Glands .... [2]

66

Coordination and response

14.3 Synapses and drugs

Language lab

Fill in the gaps.

A nerve cell is also known as a .... Impulses enter the cell body through the .... and leave the cell body along an ....

1. The diagram below shows a single synapse.

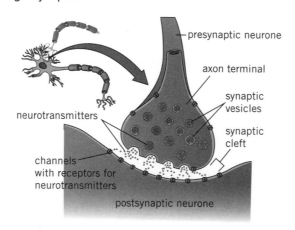

a. Use the diagram to define:

   i. Neurotransmitter ....

   .... [1]

   ii. Synapse ....

   .... [1]

b. Explain how a synapse is important in the operation of a reflex arc.

   ....

   .... [2]

c. Use the diagram to explain how drugs such as heroin and amphetamines have their effects on the body.

   ....

   .... [2]

67

Coordination and response

14.4 Sense organs

Language lab

A change in the environment which causes a reaction in a receptor is a:

A synapse

B stimulus

C mutation

D synthesis

1. The diagram shows a tennis player. The player uses different receptors during a tennis match.

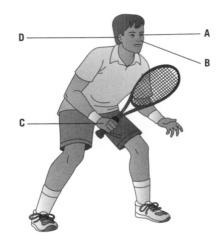

a. Complete this table to describe the function of these receptors.

letter	name of receptor/sense organ	function of receptor
A		
B	nose	
C		
D		detects position – important in balance

[6]

b. All receptors convert a stimulus into a 'message'. State the nature of this message, and briefly explain how the messages reach the central nervous system.

....

....

.... [2]

Coordination and response

14.5 The eye

Language lab

Which of the following is not a part of the mammalian eye?

A retina

B rectum

C rod

D choroid

1. The diagram shows a section through the eye.

   a. Use the table below to match up the lettered structures with the descriptions of their functions.

description	letter
contains rods and cones	
helps to converge light towards the retina	
is black to prevent internal reflection of light	
is tough enough to act as an attachment for the muscles that move the eye in its socket	
is a muscle controlling the amount of light entering the eye	
contains neurones leading to the visual centre in the brain	

   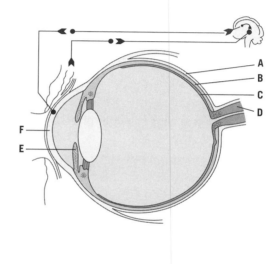

   [6]

   b. The diagram also shows the pathway of an important reflex involved in protection of the eye.

   i. Complete and rearrange the boxes to show that you understand the pathway of this action.

   receptor is ....

   effector is ....   stimulus is ....

   response is ....   coordinator is ....

   [6]

   ii. State the survival value of this reflex.

   .... [1]

69

Coordination and response

14.6 Hormones

Language lab

Which one of the following organs produces both digestive enzymes and hormones?

A adrenal gland

B ovary

C pancreas

D pituitary

1. a. Define the terms.

   Hormone ....

   .... [1]

   Target organ ....

   .... [1]

   b. State **two** ways in which control by hormones is different from control by the nervous system.

   1. ....

   ....

   2. ....

   .... [2]

Coordination and response

14.7 Controlling conditions in the body

Language lab

Fill in the gaps.

.... is the set of processes which keep constant conditions in the body.

This set of processes usually rely on .... feedback.

1. The diagram below shows a section through human skin in warm conditions.

   a. State the names of the structures labelled **A** and **B**.

   A ....

   B .... [2]

   b. i. State what happens to **B** if the erector muscle receives a stimulus to contract.

   .... [1]

   ii. Explain how this response helps to regulate body temperature in a cold environment.

   ....

   ....

   .... [2]

   iii. On the diagram (right) the blood capillary is not complete. Complete the diagram to show the size and position of the capillary in cold conditions. [2]

   c. The control of body temperature is an example of **negative feedback**.

   Explain what is meant by the term negative feedback.

   ....

   ....

   .... [2]

Coordination and response

14.8 Controlling body temperature

Language lab

Which of the following would not result from a rise in body temperature?

A   shivering   B   vasodilation

C   sweating   D   lowering hairs on the skin

1. The diagram shows a section through human skin.

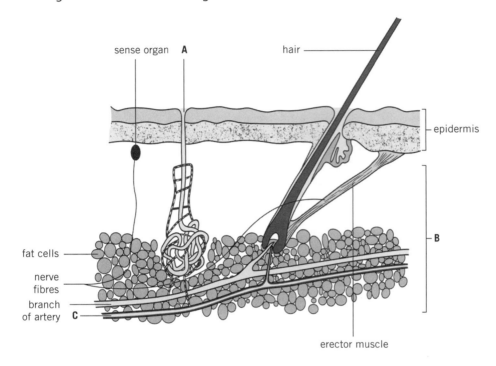

   a. State the name of the parts labelled **A**, **B**, and **C**.

   A .... B .... C .... [3]

   b. State and explain the function of the fat cells.

   ....

   ....

   .... [2]

   c. The body temperature of a mammal remains constant within narrow limits in spite of changes in the environmental temperature.

   State **two** ways in which this 'constant' body temperature is valuable to mammals.

   1. ....

   ....

   2. ....

   .... [2]

Coordination and response

14.9 Tropic responses

Language lab

The following words are scrambled versions of terms associated with plant responses. Unscramble the words.

XAUNI   TOIMSVGRAIPR   TMIOSHPTOORP

Auxin is a hormone made by the tips of plant shoots.

1. a. A shoot was grown in a container so that light shone onto it from one side only. This set of diagrams shows movement of the auxin in the shoot, and the results of the experiment.

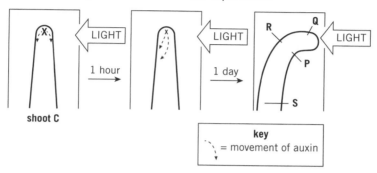

   i. Describe the movement of auxin in the shoot after one hour.

   .... [1]

   ii. Use the diagram to explain how the movement of auxin causes the plant response seen in this investigation.

   ....
   ....
   ....
   .... [2]

   iii. Name the plant response seen above, and explain its value to a plant.

   Name of response .... [1]

   Benefit to plant .... [1]

   b. Plant hormones can be used by farmers to manage plant growth.

   Give **two** examples of the commercial value of plant hormones.

   i. ....
   ....

   ii. ....
   .... [2]

73

Drugs

15.1 Drugs

Language lab

True or false?

Alcohol and heroin are powerful stimulants.

1. a. Complete the following table about some drugs. Use phrases from the following list. [5]

name of drug	effect on the human body
heroin	
nicotine	
oestrogen	
penicillin	
testosterone	

Protects against syphilis

Gives a feeling of calmness and rest

Stimulates the heart rate

Protects against AIDS

Increases growth rate in muscle

Stimulates ovulation

b. Explain what is meant by **addiction**. ....

....

.... [2]

c. Not all drugs are addictive, but overuse can still be dangerous. Explain what is meant by **antibiotic resistance**, and suggest why it might be dangerous.

....

....

.... [2]

74

Drugs

15.2 Heroin

Language lab

Complete the following paragraph.

Heroin is a powerful ...., and acts on the .... in the nervous system. Long-term use of the drug can cause ...., and a heroin user who stops taking the drug risks suffering from .... symptoms.

1. This diagram shows how some drugs affect the nervous system.

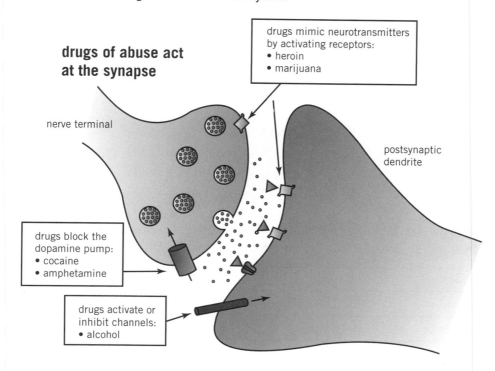

a. Use the diagram, and your biological knowledge, to suggest:

   i. why heroin has been used as a pain reliever

   ....
   ....
   .... [3]

   ii. the meaning of **drug tolerance**.

   ....
   .... [2]

b. Outline **three** social costs of heroin addiction (bullet points will be adequate for this answer).

   ....
   ....
   .... [3]

75

Drugs

15.3 Alcohol and the misuse of drugs in sport

Language lab

EPO and testosterone are two drugs used by cheating athletes. .... increases the number of red blood cells in the circulation, so helps the transport of ...., whilst testosterone increases the growth of .... which may be used for movement.

1. The diagram below shows some of the effects of drinking alcohol.

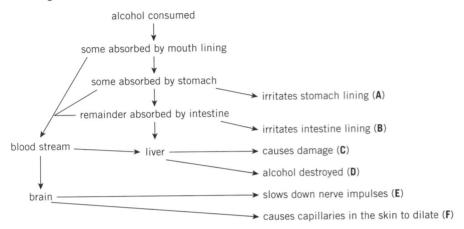

   a. Use letters from the diagram to complete the following table. The table links effects to damage caused to body processes. There may be more than one letter which corresponds to a particular example of damage.

damage to....	letter(s) of effect(s)
urea production	
digestion of protein	
absorption of fatty acids	
control of body temperature	

   [4]

   b. Alcohol may affect an unborn child.

   Explain how alcohol may reach the cells of a fetus. ....

   ....

   .... [3]

76

Drugs

15.4 Smoking and health

Language lab

Which of the following is not likely to be a result of smoking?

A cancer

B bronchitis

C cholera

D emphysema

1. a. Study this histogram.

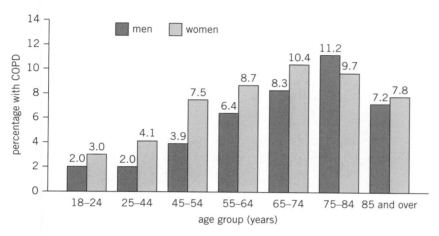

   i. Suggest a suitable title for the chart.

   .... [1]

   ii. State what the initials **COPD** stand for.

   .... [1]

   iii. Does the chart provide evidence that smoking increases the risk of COPD?

   Explain your answer. ....

   ....

   .... [2]

   b. The maximum volume of air that can be exchanged with a single breath in and out is called the vital capacity. A typical healthy male would have a vital capacity of about 5000 cm³.

   Describe and explain the likely effect of long-term smoking on vital capacity. ....

   ....

   ....

   .... [2]

Reproduction

16.1 Asexual and sexual reproduction

> **Language lab**
>
> Asexual reproduction is the process by which:
>
> A Four offspring are produced.
>
> B The offspring are genetically identical.
>
> C The offspring are genetically different.
>
> D Only one offspring is produced.

1. The diagrams below show simple schemes for asexual and sexual reproduction.

   asexual

   nn

   sexual

   nn

   a. Insert the words **meiosis** and **mitosis** in the correct positions on the diagram. [2]

   b. Write in the correct chromosome number in the boxes on the diagram. [2]

   c. Suggest one **advantage** and one **disadvantage** of asexual reproduction in plants.

   Advantage ....

   Disadvantage .... [2]

Reproduction — 16.2 Flower structure

> **Language lab**
>
> Pollen grains are formed in the:
>
> A carpel
>
> B petal
>
> C anther
>
> D filament

1. Most of the variations in flower structure are related to the methods of pollination.

    Define the term **pollination**. ....

    .... [1]

    This diagram shows a typical wind-pollinated flower.

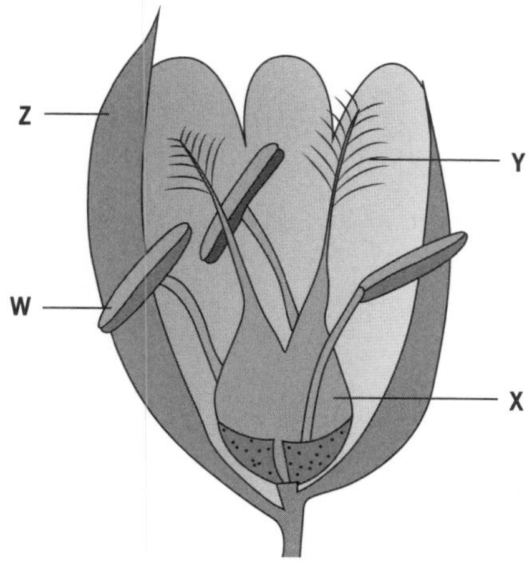

    Identify the structures labelled **W**, **X**, **Y**, and **Z**.

    W.. X..

    Y.. Z.. [4]

    Complete this table to compare the structure of wind- and insect-pollinated flowers.

part of flower	wind-pollinated	insect-pollinated	explanation
petals			
anthers			
pollen			
stigmas			

    [8]

Reproduction   16.3 Pollination

Language lab

Pollination is:

A The transfer of sperm to egg cells.

B The spreading of seeds from a plant.

C The removal of nectar by an insect.

D The transfer of pollen from stamen to stigma of a flower.

1. The diagram below shows half of a tomato flower.

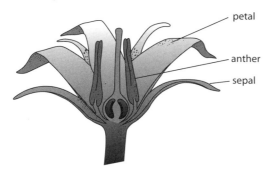

a. i. Mark with a **P** the part where a pollen grain must land to pollinate the flower. [1]

ii. State the name of the part where the pollen grain has landed.

.... [1]

b. This diagram shows a pollen grain with a pollen tube that is growing towards an ovule.

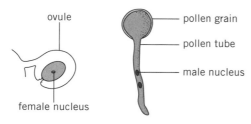

i. Complete the drawing of the pollen tube to show how it enters the ovule. [2]

ii. Describe and explain what happens to the male and female nucleus at fertilisation.

....

....

....

....

.... [3]

80

Reproduction
16.4 Fertilisation and seed formation

Language lab

Complete this paragraph by filling in the gaps.

A seed contains an ……………………… which can eventually develop into a young plant, a food store which is often the carbohydrate ……………………… The whole seed is enclosed in a protective coat called the ……………………… .

1. The diagram shows a carpel (the female part of a flower).

    a. Match the letters on the diagram with the names of structures in this table.

    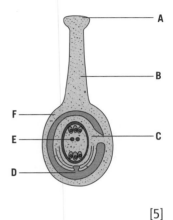

name of structure	label letter
egg cell (female gamete)	
ovule	
style	
stigma	
micropyle	
ovary	

    [5]

    b. Complete the diagram to show how a male nucleus fertilises the egg cell. Add labels to your diagram. [3]

    c. This diagram shows the appearance of a broad bean seed which has been sectioned and stained with iodine solution.

    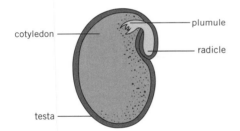

    i. State which parts of the broad bean seed make up the embryo.

    …………………………………………… ………………………………………………… [1]

    ii. State the name of the main food material stored in the broad bean seed.

    …… [1]

    iii. Explain what change you would expect to see in the appearance of the seed to make you draw this conclusion.

    …… [1]

81

Reproduction

16.5 The male reproductive system

Language lab

Which statement is true for both eggs and sperm?

A They are both motile.

B They are produced throughout life.

C They both produce hormones.

D They contain the haploid number of chromosomes.

1. This diagram shows the reproductive system of a human male.

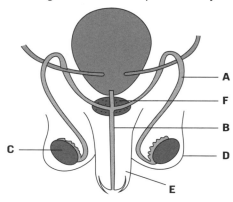

a. State the names of the parts labelled **A**, **B**, **C**, **D**, **E**, and **F**.

   Write your answers in the table below.

A	
B	
C	
D	
E	
F	

   [6]

b. Use the letters to state:

   i. the part which produces the male gametes .... [1]

   ii. the part which produces the liquid part of semen .... [1]

   iii. the part which produces testosterone .... [1]

   iv. the part which also carries urine .... [1]

   v. the part which is cut during the process of vasectomy. .... [1]

c. Arrange the following processes into the correct sequence necessary for the production of a human baby.

   A ejaculation   B implantation   C birth

   D fertilisation   E ovulation   F development

   .... [3]

82

Reproduction

16.6 The female reproductive system

Language lab

The site of gamete formation in the human female is the:

A uterus

B ovary

C oviduct

D vagina

1. This diagram shows the human female reproductive system.

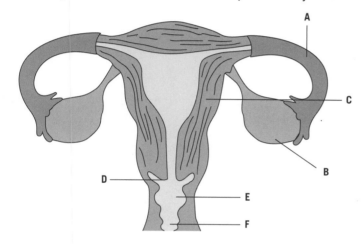

Use the letters to state:

a. the part which produces the female gametes .... [1]

b. the part where fertilisation normally occurs .... [1]

c. the part where sperm are deposited during intercourse .... [1]

d. the part where implantation occurs .... [1]

e. the part where a contraceptive diaphragm is placed. .... [1]

Reproduction — 16.7 Fertilisation and implantation

Language lab

The site of implantation in the human female is the:

A uterus

B ovary

C oviduct

D vagina

1. The diagram below shows a front view of the human female reproductive system.

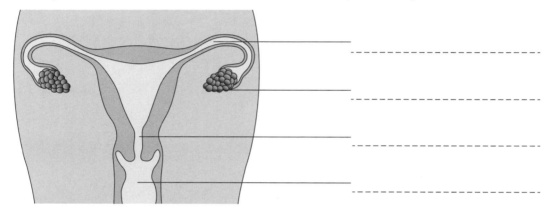

   a. Add labels to the diagram. [4]

   b. i. Name the organ where the sperm usually fertilises the egg.

   .... [1]

   ii. State the name of the structure formed when an egg is fertilised.

   .... [1]

   c. Match each of the following terms with the appropriate definition.

   AID   conception   copulation   development   fertilisation   implantation

definition	term
the fusion of male and female gametes	
the beginning of development of a new individual	
sexual intercourse	
fertilisation using sperm from a donor male	
the attachment of the fertilised egg to the lining of the uterus	
the stages which occur as cells divide and become organised into tissues and organs	

   [5]

84

Reproduction 16.8 Pregnancy

Language lab

The period of time between conception and birth is called:

A the gestation period   B the menstrual cycle

C parturition phase   D ovulation

1. The diagram below shows a human uterus containing a developing fetus.

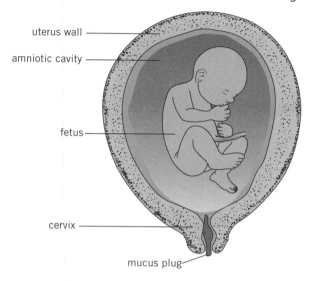

a. i. Name two structures shown in the diagram which protect the fetus.

For each structure you have named, explain how it carries out its protective function.

Structure 1: name .... [1]

Explanation of function ....

.... [1]

Structure 2: name .... [1]

Explanation of function ....

.... [1]

ii. Draw in the position of the placenta on this diagram. [1]

b. Complete this table by placing a tick (✓) in ONE box on each line to show the movement, if any, of each substance across the placenta.

substance	from mother to fetus	from fetus to mother	no movement across placenta
glucose			
haemoglobin			
nicotine			
amino acids			
carbon dioxide			
alcohol			
urea			

[7]

85

Reproduction

16.9 Antenatal care and birth

Language lab

Human breast milk gives a positive test with biuret reagent. This suggest that it contains:

A sugar

B sodium chloride

C fat

D protein

1. The table below compares the minimum daily requirements of infants (under 18 months old) and adults (over 18 years old).

	energy requirements (kJ)	protein (g)	iron (mg)	calcium (mg)	vitamin D (µg)
infant	2400	15	6	650	10
adult	8800	45	10	450	2.5

a. Calculate the percentage of the adult's energy requirement required by the infant. Show your working.

……………………………………% [2]

b. Explain why the infant requires more calcium and vitamin D than the adult, but less iron.

More calcium and vitamin D because ……………………………………………………………………

…… [1]

Less iron because ……………………………………………………………………………………………

…… [1]

This table summarises the composition of 1 kg of cow's milk.

	energy content (kJ)	protein (g)	iron (mg)	calcium (mg)	vitamin D (µg)
cow's milk	2750	35	1	1200	0.3 (summer) 0.15 (winter)

c. Calculate how much cow's milk an infant would need to satisfy its energy requirement. Show your working.

……………………………………… [2]

d. Use data from the two tables to suggest a disadvantage of feeding an infant on cow's milk.

……

…… [2]

e. Suggest a reason why cow's milk contains more vitamin D in summer than in winter.

…… [1]

Reproduction

16.10 Sex hormones

Language lab

Complete the following paragraph.

The phase of human development in which an individual becomes capable of producing mature sex cells is called .... This process is closely controlled by .... which bring about physical changes called the .... characteristics.

1. The diagram shows some of the changes associated with puberty.

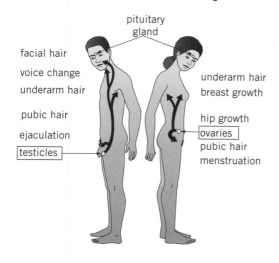

   a. i. Name the gland which first provides the signals to begin the changes associated with puberty.

   .... [1]

   ii. Name the hormone responsible for these changes in a male.

   .... [1]

   b. Suggest why the breasts grow bigger and the hips grow wider as a female passes puberty.

   Breasts ....

   Hips .... [2]

   c. Menstruation is a sign of the monthly menstrual cycles.

   State the name of the end of active menstrual cycles in the female.

   .... [1]

Reproduction

16.11 The menstrual cycle

Language lab

In a mature human female an egg cell is normally released from each ovary every .... days.

1. a. The calendar below shows the menstrual cycle for a woman in June 2011.

Monday	Tuesday	Wednesday	Thursday	Friday	Saturday	Sunday
		1	2	3	4	5
6	7	8	9	10	11	12
13	14	15	16	17	18	19
20	21	22	23	24	25	26
27	28	29	30			

key: ⊠ = ovulation (the release of an egg)   ▨ = menstruation   ▥ = when the egg is in the oviduct

   i. State how many days of menstruation are shown during June 2011.

   .... [1]

   ii. Assume that this woman has identical menstrual cycles from month to month. State the date in early July 2011 that would be the last day of menstruation.

   .... [1]

   iii. Explain why fertilisation could not occur on the 7th June 2011 even if active sperms are released into the vagina.

   .... [1]

   b. The changes in the thickness of the uterus wall during the menstrual cycle are affected by the hormones progesterone and oestrogen.

   i. State which hormone reaches its maximum 1 to 2 days before menstruation.

   .... [1]

   ii. State which hormone reaches its maximum 1 to 2 days before ovulation.

   .... [1]

Reproduction

16.12 Methods of birth control

> **Language lab**
>
> Complete the following sentence.
>
> The use of .... can be effective birth control as well as giving protection against infections such as ....

1. a. State two reasons for using a condom during sexual intercourse.

   ....

   .... [2]

   b. The table below shows some methods of contraception and their rates of failure.

method	rate of failure	ranking
condom	1 in 7	
coil or IUD	1 in 20	
diaphragm	1 in 8	
pill	1 in 300	
rhythm	1 in 4	
sterilisation	1 in 30 000	

   i. Complete the rank order in the table. The best method, i.e. lowest rate of failure, scores 1 and the one with the greatest rate of failure scores 6. [1]

   ii. Explain how the contraceptive pill prevents pregnancy.

   ....

   ....

   ....

   .... [3]

   iii. If 1000 women were using the IUD calculate how many would become pregnant after one year. Show your working.

   .... [2]

Reproduction — 16.13 Control of fertility

Language lab

Fertility may be controlled by IVF and by AI. What do these abbreviations mean?

1. The diagram below shows the changes in concentration of four hormones during the menstrual cycle.

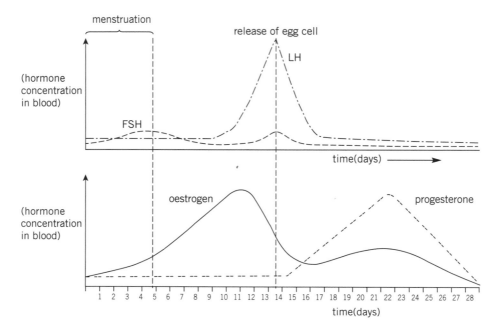

a. Oestrogen is known to inhibit the production of FSH by feedback inhibition of the action of the pituitary gland, and FSH is known to stimulate development of mature eggs in the ovary. Progesterone inhibits formation of both LH and FSH.

Use the graph, and the information above, to explain how one or more of these hormones might be used to treat infertility.

....

....

....

.... [3]

b. Suggest which hormone might be useful in preparation for IVF. Explain your answer.

.... because ....

.... [2]

c. Women who are lactating produce high levels of oestrogen. Suggest how this might act as a natural method of contraception.

....

.... [1]

Reproduction

16.14 Sexually transmitted infections (STIs)

Language lab

HIV is an organism which can be transmitted by sexual activity. This organism is harmful because it affects which type of cell?

A egg cells

B red blood cells

C white blood cells

D brain cells

1. Gonorrhoea is a sexually transmitted infection (an STI).

   a. i. State which **type** of organism causes gonorrhoea.

   .... [1]

   ii. State **one** early symptom of gonorrhoea.

   .... [1]

   iii. State which method of contraception is most likely to prevent transmission of this disease.

   .... [1]

   b. Another STI is AIDS.

   i. Explain how AIDS affects the human body.

   .... [2]

   ii. Explain why AIDS and gonorrhoea cannot be treated by the same drugs.

   .... [2]

Inheritance

17.1 Chromosomes, genes, and DNA

Language lab

A section of DNA which codes for a single protein is a:

A chromosome B nucleus

C gamete D gene

1. The diagram shows a short sequence of bases in a DNA molecule. The base sequence in DNA carries information in the form of a genetic code.

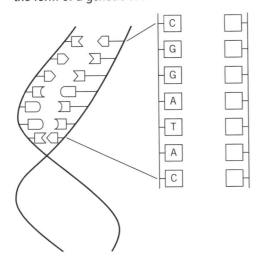

   a. Write down the sequence of bases on the other strand of the DNA molecule. [2]

   b. A modified copy of the DNA sequence is carried to the cytoplasm of the cell.

      i. State the name of this modified sequence. .... [1]

      ii. State the name of the structure in the cell where the code is 'read' by the cell. .... [1]

   c. A sequence of three bases carries a code for another type of biological molecule.

      State the name of this type of molecule. .... [1]

   d. The type of molecule named in your answer to part c. is built up into larger molecules, and these molecules determine the characteristics of organisms.

      i. State the general name given to these larger molecules. .... [1]

      ii. Complete this table to match the molecules to their functions (i.e. to the characteristics they give to an organism). [5]

molecule	characteristic
	ability of red blood cells to transport oxygen
	bind to and identify molecules on the surface of invading microbes
receptor protein in synapse	
lipase	
	provides strength and structure to hair and nails

Inheritance

17.2 Protein synthesis

Language lab

Complete the following paragraph.

Proteins are produced in organelles called ………………………… by joining together molecules called ………………………………, using instructions provided by sections of DNA called ……………………………… .

1. Scientists have found that they can use small rings of DNA called plasmids to transfer useful genes into bacterial cells. The bacterial cells can then manufacture the protein coded by this gene. The protein can be extracted from the bacterial cells, and may be very valuable in medical treatments for humans.

   a. One useful protein made by genetic engineering can be used to break down starch during food production.

      State the name of this protein. ………………………………………………………………………………………… [1]

   b. Some people believe that genetic engineering can be dangerous, but not everyone agrees with this.

      Suggest two advantages of, and two possible concerns about, genetic engineering.

      Two advantages ………

      …… [2]

      Two possible concerns ……………………………………………………………………………………………………

      …… [2]

Inheritance

17.3 Mitosis

Language lab

When a human cell with 46 chromosomes divides by mitosis it will produce:

A One cell with 46 chromosomes.

B Two cells with 23 chromosomes each.

C Two cells with 46 chromosomes each.

D Four cells with 23 chromosomes each.

1. Human body cells usually contain 23 pairs of chromosomes. The exceptions to this rule are the gametes and the mature red blood cells.

   a. Complete the table below.

type of cell	total number of chromosomes	type of sex chromosomes present
male nerve cell		
female white blood cell		
sperm cell		
egg cell/ovum		
red blood cell		

   [5]

   b. Blood cells are produced in bone marrow. State the name of the type of cell division which produces them.

   .... [1]

2. This diagram shows some stages in cell division.

   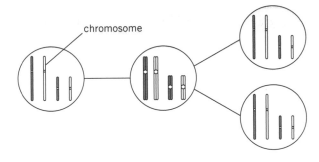

   a. State the diploid number of chromosomes in these cells. .... [1]

   b. Name the type of cell division shown. .... [1]

   c. Give **two** examples of sites where this type of cell division occurs.

   1. ....

   2. .... [2]

Inheritance

17.4 Meiosis

Language lab

Which two of the following cells do not contain a diploid nucleus?

A An egg cell and a muscle cell.

B A red blood cell and a sperm cell.

C A sperm cell and a nerve cell.

D A muscle cell and a red blood cell.

1. The photomicrograph shows a set of chromosomes during meiosis.

a. Complete the diagram below by drawing in the chromosomes of the four cells produced when the original cell divides by meiosis.

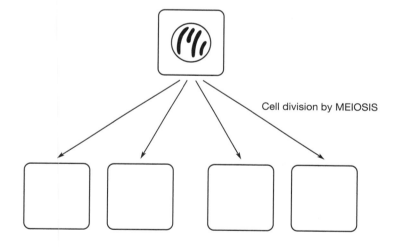

Cell division by MEIOSIS

[4]

b. State the site of meiosis in:

  i. a mammal .... [1]

  ii. a flowering plant. .... [1]

c. Explain why it is necessary for gametes to be produced by meiosis.

....

....

.... [2]

Inheritance

17.5 Inheritance and genes

Language lab

In humans the allele for brown eyes (B) is dominant to the allele for blue eyes (b). If one parent is represented as BB and the other as Bb, then they will both:

A have brown eyes but different genotypes.

B have blue eyes and the same genotypes.

C have the same phenotypes and genotypes.

D be heterozygous with brown eyes.

1. This paragraph describes some features of inheritance.

   Use words from this list to complete each of the spaces in the paragraph. Each word may be used once, or not at all.

   | allele | diploid | discontinuous | dominant | gene | haploid | heterozygous |
   | homozygous | meiosis | mitosis | recessive | redundant | | |

   In humans, eye colour may be brown or blue. This is controlled by a single ………………………….. which has two forms.

   Gametes are formed by the type of cell division called ………………………………: the gametes are

   ……………………………………….. and fuse at …………………………………………….. to form a

   ……………………………………. zygote.

   Two humans both have brown eyes, but one of their three children has blue eyes. This means that blue eye colour is

   controlled by a ……………………………….. allele and that both of the parents are ………………………………………. [7]

2. Study the two lists below. One is a list of genetic terms and the other is a list of definitions of these terms.

   Draw guidelines to link each term with its correct definition.

genetic term		definition
genotype		the observable features of an organism
homozygous		an allele that is always expressed if it is present
dominant		having two alternative alleles of a gene
heterozygous		one alternative form of a gene
recessive		the set of alleles present in an organism
chromosome		having two identical alleles
allele		a thread-like structure of DNA, carrying genetic information in the form of genes
phenotype		two identical alleles of a particular gene

   [8]

Inheritance — 17.6 Monohybrid inheritance

Language lab

A couple who are both carriers (heterozygous) for a fatal genetic condition might be advised not to have children because:

A all of their children will show the condition.

B the child's genes are more likely to mutate.

C the child could receive the recessive allele from each parent.

D the mother could be harmed during the pregnancy.

1. Tongue rolling is an example of discontinuous variation. It is partly controlled by a dominant allele, R, of a single gene.

   a. i. Name one other example of discontinuous variation in humans.

      .... [1]

      ii. Define the term *allele*.

      .... [1]

      iii. Draw a simple labelled diagram to show how genes and chromosomes are related to one another. [2]

   b. This diagram shows a family history (a pedigree) of tongue rolling over three generations.

   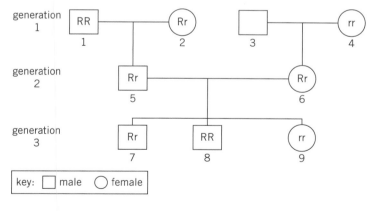

   key: ☐ male ◯ female

      i. State which individual in generation 3 is *not* a tongue roller. .... [1]

      ii. Suggest the possible phenotypes of individual 3. .... [1]

      iii. Individual 7 marries a woman who is heterozygous for tongue rolling.

         State the probability (chance) that their second child will be a tongue roller.

         .... [1]

Inheritance 17.7 Codominance

Language lab

In one family, there were four children with different blood groups, i.e. A, B, AB, and O. Which one of the following combinations were the parent's genotypes?

A AO and BO

B AB and OO

C AB and AB

D AO and AB

1. a. A set of triplets was born, but their mother died during the birth. The babies were separated from one another, and brought up by different families. When they were 18 they met up with one another for the first time since their separation. Various measurements were made on them, and some of the information obtained is recorded in this table.

	Andrew	John	David
mass (kg)	81	92	86
height (cm)	179	180	179
blood group	O	O	AB
intelligence quotient (IQ)	128	138	140

   i. State which two of the boys might be identical twins from this evidence.

   .... and .... [1]

   ii. Explain which piece of evidence is most important in helping you to reach this conclusion.

   ....

   .... [1]

   iii. Suggest a reason why Andrew and David have such different body mass measurements although they are of the same height.

   .... [1]

   b. The ABO blood group is determined by three alleles, although only two are present in any one cell. The three alleles are given the symbols I^A, I^B, and I^O.

   The relationship between genotype and phenotype for these blood groups is shown in the table below.

genotype	$I^A I^A$	$I^A I^O$	$I^B I^B$	$I^B I^O$	$I^A I^B$	$I^O I^O$
phenotype	A	A	B	B	AB	O

   State the name of this type of genetic relationship, in which a particular combination of alleles (in this case $I^A I^B$) results in a new phenotype.

   .... [1]

Inheritance

17.8 Sex linkage

> **Language lab**
>
> N is the allele for normal colour vision, and n is the allele for red-green colour blindness. An individual with the genotype $X^N X^n$ is:
>
> A a boy with normal colour vision
>
> B a colour blind boy
>
> C a colour blind girl
>
> D a girl who is a carrier for colour blindness

1. This diagram shows the arrangement of chromosomes in a cell of a human fetus.

    a. State the gender (sex) of this fetus. .... [1]

    Explain your answer. .... [1]

    b. The fetus has a chromosome mutation, which may lead to a genetic abnormality. Use the diagram to identify this chromosome mutation.

    .... [1]

2. Haemophilia is a sex-linked inherited disease. It is controlled by alleles carried on the X chromosome (**H** for normal clotting, **h** for haemophilia).

    The table below gives some information about possible combinations of phenotype and genotype.

    Complete the table by:

    a. drawing the correct chromosomes in boxes 2, 3, and 5

    b. adding the symbols for the alleles to the correct positions on the chromosomes

    c. writing into the final row the sex of each individual.

	possible combinations				
	1	2	3	4	5
diagram	H—⫯ ⫯—H			H—⫯ ⫯	
sex chromosomes	XX	XX	XX	XY	XY
sex of person					

    [5]

99

Variation and selection   18.1 Variation

Language lab

Complete the following paragraph.

The sum of the characteristics of an organism is called its ...., and results from a combination of its set of genes (its ....) and the effects of the ....

1. Study the two lists below. One is a list of terms relating to variation and the other is a list of definitions of these terms.

   Draw guidelines to link each term with its correct definition.

genetic term
continuous variation
gene
discontinuous variation
phenotype
height in humans
environment
nutrients
blood group

definition
the observable features of an organism
a form of variation with many intermediate forms between the extremes
one possible form of environmental influence on variation
one example of discontinuous variation
this factor, in addition to genotype, can affect phenotype
a section of DNA responsible for an inherited characteristic
one example of continuous variation
a form of variation with clear-cut differences between groups

[8]

Variation and selection — 18.2 Mutations

Language lab

Which of the following is unlikely to act as a mutagen?

A ultraviolet light B X-radiation

C excess vitamin C D tar from cigarette smoke

---

1. a. Sickle cell anaemia is inherited. It is caused by a recessive allele (Hb_S): the normal red blood cell allele is Hb_A.

   Study the family tree (below).

   key:
   - ○ female not affected
   - ◉ female carrier
   - ? female, condition unknown
   - □ male not affected
   - ■ male carrier
   - ? male, condition unknown

   i. State Lucy's genotype. .... [1]

   ii. Suggest which person in the family tree could have sickle cell anaemia. Explain your answer.

   Name ....

   Explanation ....

   .... [2]

   b. A person heterozygous for sickle cell anaemia (Hb_A Hb_S) shows some symptoms of anaemia. These may lead to weakness and tiredness, although they are rarely severe enough to be fatal. Despite these disadvantages there are situations in which the sickle cell allele may give an advantage to the carrier.

   Explain how the Hb_S allele may be a selective advantage in certain environments.

   ....

   ....

   .... [2]

2. Charles Darwin believed that new **species** can arise by natural selection of variations within a population. Since Darwin's time, scientists have suggested that variation may arise by **mutation**, as well as by the formation of new combinations of **genes** at fertilisation during sexual reproduction.

   a. State the meanings of the terms:

   i. species .... [1]

   ii. mutation .... [1]

   iii. gene .... [1]

   b. Suggest **two** factors that may increase the rate of mutation in a population.

   i. ....

   ii. .... [2]

Variation and selection   18.3 Adaptive features

> **Language lab**
>
> Complete the following sentence.
>
> .... is a change in structure of an organism to suit it to its environment, and results from an alteration in an organism's .... (its set of genes).

1. a. The table below shows some examples of selection.

example	type of selection: natural or artificial
Cattle being bred that are able to withstand cold winters.	
The development of a strain of rice that is more resistant to disease.	
The resistance of a strain of bacteria to a particular antibiotic.	
The ability of a species of shrub to grow on soil containing high amounts of copper.	
The similarity between a moth's wing pattern and its habitat, making it less conspicuous to predatory birds.	

   Complete the table to show which type of selection, natural, or artificial, is involved. [5]

   b. The bar graph shows the average milk yield for a herd of Friesian cows between 1976 and 2004.

   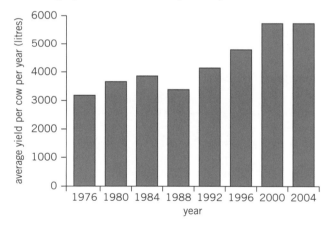

   i. Explain how animal breeders might have caused the trend between 1976 and 2004 to occur.

   ....

   ....

   .... [3]

   ii. Apart from the yield of food (milk or meat), suggest another characteristic that farmers might improve in animals to make a farm more economically successful.

   .... [1]

Variation and selection 18.4 Natural selection

Language lab

The dark form of the peppered moth lives mainly in industrial areas because it:

A is a mutant

B is well camouflaged from its natural predators

C can only find suitable breeding sites in such areas

D can find food more easily in industrial areas

1. The Everglades is an enormous flooded area in southern Florida, USA.

    There are many different species of animals and plants there, and they are well adapted to their environment.

    a. The animals and plants have become well adapted by the process of natural selection.

       The following are stages in the evolution of species by means of natural selection.

       A survival of the fittest

       B over-production of offspring

       C competition causes a struggle for existence

       D advantageous characteristics are passed on to offspring

       E variation occurs between members of the same population

       Use the letters **A–E** to rearrange these stages into the correct sequence.

       ….. ….. ….. ….. ….. [5]

    b. Some of the early European explorers in Florida noticed that there were many species of bird.

       The heads and beaks of birds are often well adapted to their diet. The diagrams in this table show the heads of some species of birds.

       Draw guidelines to match up the heads with the likely diet of the birds.

| Feeds on nuts and other hard fruits | Filters algae and other small organisms from the water | Feeds by spearing fish and frogs | Captures Florida rabbits and other mammals | Feeds by catching small insects |

[5]

103

Variation and selection — 18.5 Selective breeding

Language lab

Complete the following paragraph.

All spaniels belong to the same .... but may be very different in appearance as a result of

....

1. a. The diagrams below show the effect of selective breeding on wild cabbage (*Brassica oleracea*).

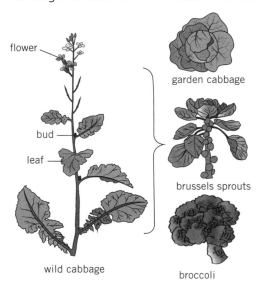

State which part of the wild cabbage has been selected for when breeding the:

brussels sprouts ....

broccoli. .... [2]

b. Once a plant has been selected as a new variety of broccoli, the grower then produces many more identical plants by taking cuttings.

i. Explain why the cuttings grow to be identical to the parent plant.

....

....

.... [2]

ii. The cuttings are usually planted into a type of compost containing plant hormones (to encourage rooting) and additional nitrate. The cuttings are kept in a closed transparent enclosure.

Explain the importance of:

nitrate in the growth compost ....

.... [1]

keeping the plants in a closed environment. ....

.... [1]

Organisms and their environment

19.1 Energy flow

Language lab

The energy flowing through an ecosystem arrives from:

A combustion of fuels   B solar energy from the Sun

C nuclear power stations   D geothermal energy from volcanoes

1. a. This diagram represents the energy flow through a food chain.

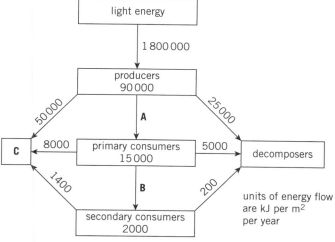

i. Name the source of light energy. .... [1]

ii. State the process occurring at **A** and **B**. .... [1]

iii. Explain why not all of the light energy available is absorbed by the producers.

   ....

   .... [2]

iv. State the name of the process shown at **C**, and the form in which the energy is lost.

   Process .... [1]

   Form of energy lost .... [1]

b. i. Calculate the percentage of energy transferred at stages **A** and **B**.

   Show your working.

   A ....   B .... [2]

ii. Suggest how evidence in the diagram supports the idea that humans should eat more vegetable matter and less meat.

   ....

   ....

   ....

   .... [3]

105

Organisms and their environment

19.2 Pyramids of numbers and biomass

Language lab

Complete the following sentence.

In producing a pyramid of biomass, the biomass is calculated by multiplying .... and ....

1. Feeding relationships.

   Use guide lines to match the terms with their definitions.

term/process		definition
food chain		an organism that gets its energy from dead or waste organic material
food web		an organism that gets its energy by feeding on other organisms
producer		an animal that gets its energy by eating other animals
consumer		a network of interconnected food chains
herbivore		the transfer of energy from one organism to the next, beginning with a producer
carnivore		an organism that makes its own organic nutrients, usually through photosynthesis
decomposer		an animal that gets its energy from eating plants

2. A group of students studied a food web for an English lake.

   Algae are small microscopic plants. Water fleas are small crustaceans about 2 mm in length. Smelt are fish – they are adult at about 4 cm in length. The kingfisher is a bird, about 22 cm long.

   Draw and label a likely pyramid of numbers for the food chain linking:

   algae ⟶ water fleas ⟶ smelt ⟶ kingfisher

   [2]

Organisms and their environment

19.3 Shortening the food chain

Language lab

The energy content of a piece of biological material can be measured in:

A   kilojoules   B   kilograms

C   degrees   D   watts

1. The image shows cattle feeding on dried pellet food.

The pellets of food are made by collecting seeds and grains from crop plants, and drying them.

Humans eat some of the cattle.

a. Explain why it would be more efficient if humans ate the grains and seeds instead of meat from the cattle.

....

....

....

.... [3]

b. The cattle are often kept inside heated sheds. Explain why this might save money for the farmer.

....

....

....

.... [3]

107

Organisms and their environment

19.4 Nutrient cycles

> **Language lab**
>
> Fill in the gaps.
>
> The simplest carbon cycle would involve an approximate balance between two processes,
>
> .... and ....

1. The first part of this question asks you to use a list of words to fill in a set of boxes. This is quite a common way to test your knowledge and understanding of a biological process. The diagram below represents the carbon cycle.

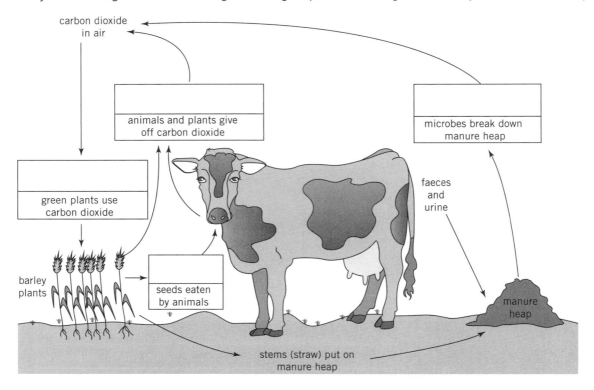

   a. Use words from this list to complete the boxes in the diagram.

   combustion   decay   excretion   feeding   photosynthesis   respiration   [4]

   b. In some countries the manure heap is collected, dried, and then burned as a fuel.

   State the effect that this would have on the carbon dioxide concentration in the air.

   .... [1]

108

Organisms and their environment

19.5 The nitrogen cycle

Language lab

Complete the following sentence.

During the process of denitirification, ... is changed to .... by ....

1. a. The diagram shows the nitrogen cycle.

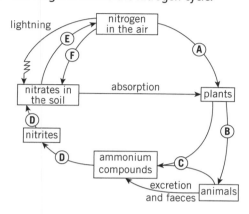

   i. Name the processes **B** and **E**.

   B ....

   E .... [2]

   ii. Name the process and the organisms involved in **C**.

   Process ....

   Organisms .... [2]

b. A heap of dead leaves had been left to decompose. Part of the decomposition sequence is shown below.

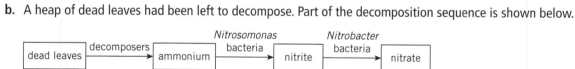

Use this sequence and your own knowledge to answer the following questions.

   i. Suggest **three** effects on the sequence if the *Nitrosomonas* bacteria died out.

   1. ....

   2. ....

   3. .... [3]

   ii. Tick **two** boxes in this table to show factors which would speed up the decomposition process.

factor	box
maintaining an acid pH	
a plastic cover to exclude air	
few scavenging insects	
many scavenging insects	
regular turning of the heap to add air	

[2]

109

Organisms and their environment

19.6 Populations, communities, and ecosystems

Language lab

Arrange the following phases of a population curve into the correct sequence:
lag – decline – stationary – exponential

1. This diagram shows a number of factors that can affect the size of a population.

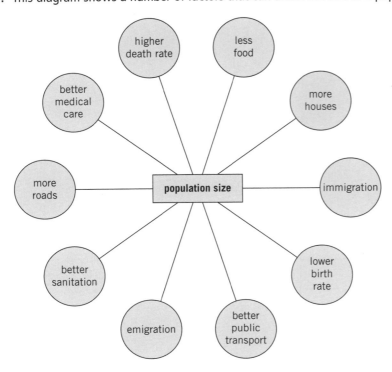

   a. State **four** factors, shown in the diagram, that reduce the size of a population.

   ....
   ....
   ....
   .... [4]

   b. Suggest **two** ways in which better medical care can lead to an increase in the size of a population.

   ....
   .... [2]

   c. State **two** ways in which humans control bacterial populations. [1]

   ....
   .... [2]

110

Organisms and their environment

19.7 Human populations

Language lab

Three 'revolutions' have contributed to the human population. In the correct order, these are the .... , ... , and the .... revolution.

1. a. Explain what is meant by the term **age structure** of a population.

   ....
   ....
   .... [2]

   b. List and explain **three** factors that can affect the age structure of a population.

   1. Factor .... **Explanation of effect** ....

   ....
   ....

   2. Factor .... **Explanation of effect** ....

   ....
   ....

   3. Factor .... **Explanation of effect** ....

   ....
   .... [6]

Biotechnology and genetic engineering

20.1 Microorganisms and biotechnology

Language lab

A small piece of DNA that can be moved from one bacterial cell to another is a:

A gene B enzyme

C ribosome D plasmid

1. a. For each of the following statements, mark as True (T) or False (F).

	Statement	True or False
1	A typical bacterium is about one thousandth of a metre wide	
2	Bacteria are smaller than viruses	
3	Gonorrhoea is caused by a bacterium	
4	Bacteria may contain plasmids	
5	All bacteria are harmful	
6	Bacteria multiply by binary fission	
7	A bacterium called *Vibrio* causes cholera	
8	Bacteria can produce the enzyme protease	
9	Bacteria can be genetically modified to produce human insulin	
10	Some bacteria can produce their own organic chemicals using energy from the Sun	
11	Bacteria in the human intestine produce some of the vitamins required for human health	
12	Bacteria have a nucleus, but it is smaller than a human nucleus	

[12]

b. For any three of the statements which you have labelled as False, explain why you have made that choice.

1. Statement ....

   Explanation ....

   .... [2]

2. Statement ....

   Explanation ....

   .... [2]

3. Statement ....

   Explanation ....

   .... [2]

Biotechnology and genetic engineering

20.2 Enzymes and biotechnology

Language lab

An important enzyme involved in clearing fruit juices is ....

1. The diagram shows a piece of equipment used in one commercial use of enzymes.

   Some enzymes are **immobilised** on a carrier of fibres. They can be re-used many times eg. lactase for lactose-free milk

   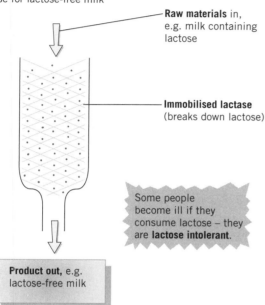

   Raw materials in, e.g. milk containing lactose

   Immobilised lactase (breaks down lactose)

   Some people become ill if they consume lactose – they are **lactose intolerant**.

   Product out, e.g. lactose-free milk

   a. Complete the diagram using words from this list.

   amylase   lactase   milk containing Lactose   lactose-free milk   starch

   [3]

   b. The enzyme is **immobilised** in this apparatus. Explain the meaning of the term immobilisation, and suggest why it is important in this process.

   Meaning of term ....

   Importance .... [2]

113

Biotechnology and genetic engineering
20.3 Fermenters

Language lab

Which one of the following is normally used to produce antibiotics?

A bacteria   B fungi

C protoctists   D viruses

1. a. This graph shows the amounts of *Penicillium* fungus and penicillin in the bioreactor over a period of nine days.

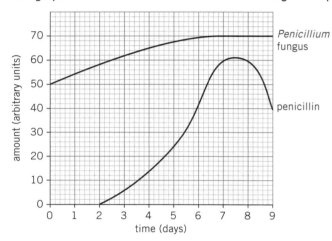

Suggest the best time to collect the penicillin.

....

Explain your answer.

....

.... [2]

b. Doctors are reluctant to prescribe penicillin for all illnesses.

Explain how the overuse of antibiotics can lead to the development of resistant strains of bacteria.

....

....

....

....

.... [3]

c. Bioreactors may also be used in the production of enzymes.

Complete this table about commercially valuable enzymes. One section has been completed for you.

name of enzyme	commercial value
	clearing fruit juices by digesting clumps of plant cells
lipase from fungi	improves chocolate flow when coating biscuits
	part of biological washing powders – remove blood stains
lactase	

Biotechnology and genetic engineering

20.4 Genetic engineering

Language lab

A product of genetic engineering, valuable to sufferers from diabetes, is:

A statin B insulin

C amylase D testosterone

1. a. The following table gives a list of some terms associated with genetic engineering, and some definitions for these terms.

   Match the terms in column 1 with the correct definition from column 2.

term
gene
plasmid
vector
ligase
restriction
sticky ends
fermenter

definition
a small circle of DNA in a bacterial cell
an enzyme that can splice one gene into another section of DNA
an enzyme that can cut out a specific gene from a chromosome
pieces of single-stranded DNA left exposed after a gene is removed from a chromosome
a section of DNA coding for a protein
a vessel in which engineered bacteria can produce a valuable product under optimum conditions
a structure which can carry a gene into another cell

   [6]

   b. Genetic engineering can be used to produce protein products that are useful to humans.

   State **one** reason why each of the following proteins is useful to humans.

   1. Insulin ....

   ....

   2. Factor 8 ....

   ....

   3. Pectinase ....

   ....

   4. Human growth hormone ....

   .... [4]

   c. Before the development of genetic engineering, vaccines were made by heating or chemically treating whole viruses. The vaccine therefore contained whole virus particles.

   Explain why it is safer to use a genetically engineered vaccine rather than one made directly from the hepatitis B virus.

   ....

   ....

   .... [2]

115

Human influences on ecosystems

21.1 Food supply

Language lab

Which one of the following is not a major source of dietary carbohydrate?

A potatoes

B rice

C wheat

D milk

1. a. This set of pie charts shows the food groups present in eggs, milk, rice, and beans.

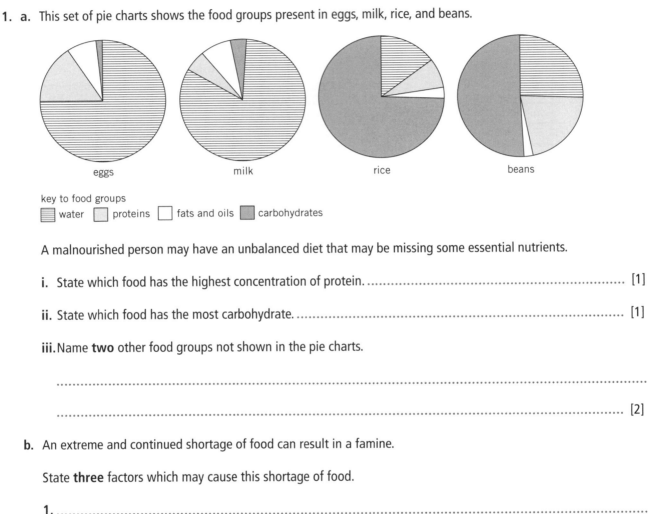

A malnourished person may have an unbalanced diet that may be missing some essential nutrients.

i. State which food has the highest concentration of protein. .... [1]

ii. State which food has the most carbohydrate. .... [1]

iii. Name **two** other food groups not shown in the pie charts.

    ....

    .... [2]

b. An extreme and continued shortage of food can result in a famine.

State **three** factors which may cause this shortage of food.

1. ....

2. ....

3. .... [3]

Human influences on ecosystems

21.2 Habitat destruction

Language lab

Fill in the gaps.

Deforestation reduces the availability of .... for many birds and increases the risk of .... erosion.

1. More food and raw materials are required to meet the demands of an increasing world population.

   Describe and explain the possible harmful effects on organisms and the environment of burning large areas of Indonesian rainforest, so that the land can be used to grow palm oil trees.

   ....

   ....

   .... [4]

2. These diagrams show changes to a farm between 1953 and 2003.

   key
   hedges ——— boundary -----
   river ～～ buildings ■■
   trees 🌳🌳 marsh ⁎⁎⁎

   The fields on the farm are separated by hedges.

   a. i. State **two** major changes which were made to the land between 1953 and 2003.

      1. ....

         ....

      2. ....

         .... [2]

   ii. Suggest and explain **two** ways in which these changes would affect wildlife on the farm.

      1. ....

         ....

      2. ....

         .... [4]

   b. Farmers often remove areas of woodland to provide more space for growing crops.

      Suggest three disadvantages of this deforestation.

      1. ....

      2. ....

      3. .... [3]

Human influences on ecosystems

21.3 Pollution

> **Language lab**
>
> Two pollutants produced by car engines are:
>
> A sulfur dioxide and nitrogen oxides
>
> B oil and water
>
> C methane and carbon dioxide
>
> D sulfur dioxide and methane

1. The table below provides some data about a number of atmospheric gases.

name of gas	source(s) of gas	percentage influence on greenhouse effect
CFCs	air conditioning systems; aerosol propellant; refrigerators	13
carbon dioxide	burning forest trees; burning fossil fuels; production of cement	56
nitrous oxide	breakdown of fertilisers	6
methane	waste gases from animals such as sheep, cattle, and termites; rotting vegetation	25

   a. Explain why these gases are called 'greenhouse' gases.

   ....

   ....

   .... [2]

   b. From this data suggest why the following practices should be encouraged:

   1. Development of renewable energy resources such as wind turbines.

   .... [1]

   2. Improved insulation of walls and roofs in houses.

   .... [1]

   3. Reforestation.

   .... [1]

2. There are large copper mines in Tanzania. One of these mines is so far from railway links that all supplies must be brought in, and all products and waste materials removed, by trucks. These trucks run throughout the day and night.

   Describe and explain the possible harmful effects on organisms and the environment.

   ....

   ....

   .... [3]

Human influences on ecosystems

21.4 Water pollution

Inorganic fertilisers can be harmful to living organisms because they:

A   are washed into rivers and pollute the water.

B   remain in the soil and poison soil animals.

C   decompose and pollute the air.

D   can be absorbed through the skin of animals.

1. Organisms in a lake or river can be affected by pollution. The statements below describe some of the effects on organisms of sewage being allowed to run into the water.

   The statements are in the wrong order.

statement	letter
Plants on the bottom of the lake die	A
Algae near the surface of the lake absorb nitrates and grow in large numbers	B
Sewage flows into the lake from a nearby farm	C
Increased algae prevent light from reaching plants rooted at the bottom of the lake	D
Bacteria break down the sewage into nitrates	E

   a. Rearrange the letters to show the correct sequence of events. The first one has been done for you.

sequence of events	first step	second step	third step	fourth step	fifth step
letter	C				

   [4]

   b. Suggest one other source of nitrates which may enter the lake.

   .... [1]

   c. The sequence that you have described in your answer to part **a.** is not the end of the harmful events that may take place.

   Describe three further steps to show how aerobic bacteria may cause the death of fish and larger invertebrates in the lake.

   1. ....

   2. ....

   3. .... [3]

Human influences on ecosystems

21.5 The greenhouse effect

> **Language lab**
>
> The two most significant greenhouse gases are .... and ....

1. Biotechnology has made it possible to use algae as a source of fuel.

   The algae are grown in a vessel called a biocoil, then dried and ground up to form a powder, which can be used as fuel.

   The biocoil is a transparent tube about 5 m high, and the algae constantly circulate through it in a nutrient solution. The algae grow and multiply at a very fast rate.

   The diagram shows the main stages in this process.

   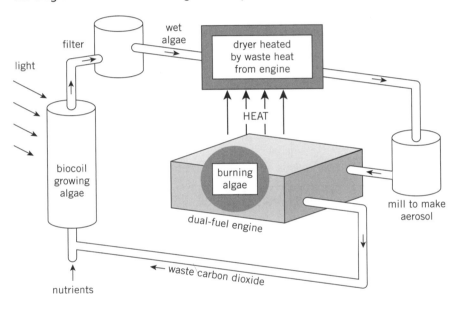

   a. Write out an equation to summarise how energy is trapped by the algae. [2]

   b. Explain why it is essential that the biocoil is transparent.

   ....
   ....
   .... [2]

   c. Explain why it is essential to use an organism with a high rate of reproduction in this process.

   ....
   ....
   .... [2]

Human influences on ecosystems

21.6 Acid rain

Language lab

Fill in the gaps.

Acid rain .... the pH of soil and water, and so increases the loss of minerals by ....

1. The diagram below shows one cause of acid rain.

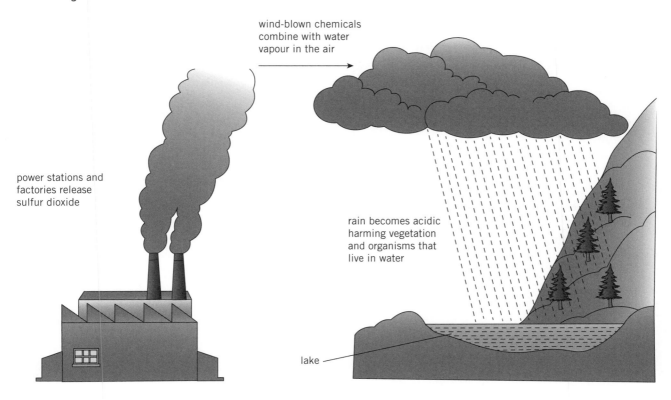

a. Describe **two** effects of acid rain on forests.

   1. ....

   2. .... [2]

b. Name **one** cause of acid rain other than that shown in the diagram.

   .... [1]

c. Suggest **two** different ways to reduce pollution so that there is less acid rain.

   1. ....

   2. .... [2]

d. Describe how acid rain can be directly harmful to **human** health.

   .... [1]

Human influences on ecosystems

21.7 Sustainable resources

Language lab

In an area where starchy food is plentiful but there is little meat or fish, farmers might be advised to grow:

A cereals   B potatoes

C legumes   D citrus fruits

1. a. Modern fishing boats use ultrasound equipment to locate shoals of fish, and giant nets with very small mesh to capture their prey.

   Describe and explain the possible harmful effects on organisms and the environment.

   ....

   ....

   ....

   .... [3]

   b. Until recent times, fishing in the North Sea was largely unrestricted.

   Suggest **three** measures that could be taken to help to conserve the cod population.

   1. ....

   ....

   2. ....

   ....

   3. ....

   .... [3]

Human influences on ecosystems

21.8 Sewage treatment and recycling

Language lab

The final stage in the treatment of sewage for safe water is to add .... which limits the growth of bacteria.

1. This diagram shows one type of sewage treatment works.

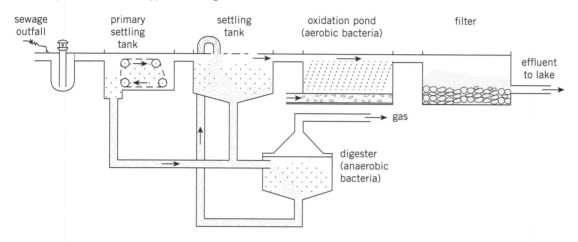

a. Define the term *anaerobic*.

   .... [1]

b. i. Name **one** gas produced by the microorganisms in the digester. ....

   .... [1]

   ii. Suggest **one** possible use of this gas.

   .... [1]

c. Explain why it is important to treat sewage before it is released into the lake.

   ....

   ....

   .... [2]

d. Processed sludge from the sewage treatment works can be used in agriculture. Describe **one** important agricultural use for the processed sludge.

   ....

   .... [2]

e. Leakage of disinfectant into the sewers can be dangerous. Explain why disinfectants in sewers can cause harm.

   ....

   ....

   .... [2]

Human influences on ecosystems

21.9 Endangered species

> **Language lab**
>
> Fill in the gaps.
>
> A species may become endangered by loss of ...., by hunting, and by failure to .... quickly enough (to replace individuals that die).

1. Define the term *conservation*. ....
   .... [2]

2. a. The Giant Panda (*Ailuropoda melanoleuca*) is an endangered species.

   i. Suggest and explain **two** reasons why the panda is endangered. One reason should relate to the environment, and one reason should relate to the biology of the panda itself.

   Environmental factor .... Explanation of effect ....
   .... [2]

   Biological factor .... Explanation of effect ....
   .... [2]

   ii. The Giant Panda is sometimes called a **flagship species**. Explain what is meant by a flagship species. ....
   ....
   .... [2]

   b. A conservation management plan involves several steps. The first step is to sample the population of the endangered species.

   i. Explain why sampling is necessary. ....
   .... [2]

   ii. For any named species, suggest a suitable method of sampling.

   Name of species ....
   Method of sampling ....
   .... [2]

Human influences on ecosystems

21.10 Conservation

Language lab

Fill in the gaps.

Conservation maintains .... (the number of different species in a habitat).

It will involve careful management by ....

1. A farmer believes in sustainable farming.

   a. What is meant by the term **sustainable**? ....

   .... [1]

   The farmer plants a forest of young fir trees. Ten years later the forest is thinned by removing some of the trees. A small part of the forest is harvested in each of the following years.

   b. Suggest **two** advantages of thinning the forest.

   ....

   .... [2]

   c. Suggest two disadvantages of growing trees as a monoculture.

   ....

   .... [2]

   d. This list contains some properties of trees. A farmer buying young trees would consider these properties when making his purchase.

   fast growth   **good growth on poor soil**   **high quality of wood pulp**

   low cost   **resistance to disease**

   Choose any **three** of these properties and suggest a reason why each property would be useful to the farmer.

property	reason for choice

125

Language focus
Helping you to use the right words

Look at the command words section (page 130).

You will see there that you might be asked to define a biological word or term. It is important that you can do this accurately, and without confusion. There are several ways to practise matching words and definitions. Some examples are given below.

Define: the answer is a formal definition of a particular term. The answer is usually 'what is it' – for example, define the term active transport means 'what is active transport'.

Look carefully how many marks are offered. It is often a good idea to add an example to a definition – in this way the examiner can be sure that you know what you are trying to define.

Matching pairs: Organisms and the environment

Use guidelines to match the terms in the first column with the definitions in the second column.

term	definition
food chain	an organism that gets its energy by feeding on other organisms
producer	a group of organism of one species, living in the same area at the same time
carnivore	an organism that gets its energy from dead or waste organic material
trophic level	an animal that gets its energy by eating plants
food web	an animal that gets its energy by eating other animals
consumer	the transfer of energy from one organism to the next
herbivore	an organism that makes its own organic nutrients, usually using energy from sunlight
decomposer	the position of an organism in a food chain
community	a unit made up of the community of organisms interacting with their non-living environment
population	all of the populations of different species in an ecosystem
ecosystem	a network of interconnected food webs

Language focus
Helping you to use the right words

Word scrambles: read the definition to help unscramble the appropriate word.

Reproduction, menstrual cycle, and pregnancy

UXESAL ODURPCERTONI: Producing new individuals with a combination of features from two parents

MEAGTES: Special sex cells, the sperm and the egg

OYTGEZ: Formed when a sperm and an egg combine

TFIRLISEITOAN: The joining together of sperm and egg

BUPETRY: A stage of human development at which the person becomes able to reproduce

ESX RHOMENOS: Chemicals that control the physical and mental changes at puberty

TEESTS: Where the male gametes are made

ROIEVAS: Where the female gametes are made

IVNAGA: The birth canal

NISEP: Delivers sperm to the vagina

CENPTOICON: The beginning of the development of a new individual

ULPATOIONC: Sexual intercourse – when the sperm from the male are delivered to the female's reproductive system

REYMOB: A stage of development when the ball of cells begins to rearrange itself so that some organs can be seen

STRNAEUTIOMN: The release of the bloody lining of the uterus if no fertilisation has taken place

WGHROT: Getting bigger by the production of more cells

EDLEEOPMVNT: The changes of cells that mean some of them take on different functions

CPLENTAA: A structure linking the umbilical cord to the wall of the uterus

TGATESNIO IPEODR: The length of time between fertilisation and birth

IULIBCALM ROCD: The structure that links the developing fetus to the placenta

INMOATCI IFDUL: Liquid inside a sac that surrounds the developing fetus

Language focus
Helping you to use the right words

Crosswords: the 'clues' are the definitions, and the 'answers' are the biological terms. Crosswords are really helpful to vocabulary training – in order to make the answers fit the grid, the spelling must be correct.

The importance of DNA

DNA is the molecule which carries the coded information responsible for all of the characteristics of organisms. The structure of this molecule was not worked out until 1953, and an entire new branch of Biology – Molecular Biology – has developed since that discovery. This exercise will remind you of some of the key features of this molecule.

Across:
1 makes up the backbone of the DNA molecule
3 the twisted ladder of DNA (6, 5)
5 found in the nucleus and made up of DNA
6 a section of DNA that codes for a single protein
7 two scientists who suggested the structure of DNA (6, 3, 5)
8 Moves substances across the cell membrane (7, 7)
13 control centre of the cell – this is where you would look for DNA
14 …. or DNA for short!
15 one of the subunits that make up DNA
16 one of the molecules that make up the rungs of the DNA ladder
17 female scientist who helped in understanding DNA structure

Down:
2 a protein found in red blood cells
4 a rule that 'links' the two chains of the DNA molecule (4–7)
9 the copying of DNA
10 a molecule responsible for one characteristic of a cell
11 type of cell division that produces exact copies
12 protein molecule found in hair cells

Language focus
Helping you to use the right words

Controlling inheritance

Once scientists understood the processes involved in reproduction and inheritance, they tried to control these processes for the benefit of humans. This higher level exercise tests your recall of some of the key terms used to describe this control.

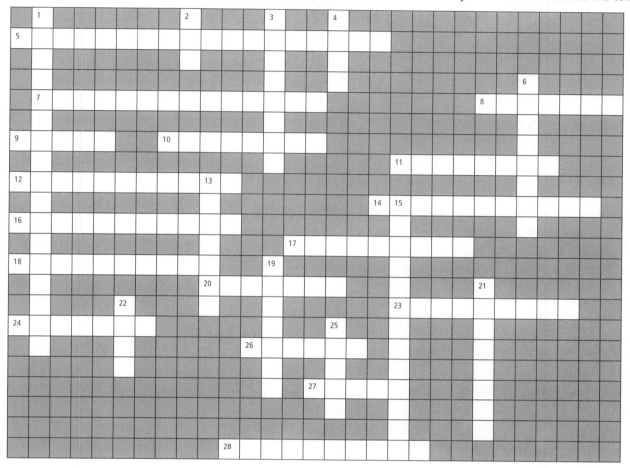

Across:

- 5 techniques for changing an organism's genotype (7, 11)
- 7 a disease of the lungs that may be treated by 12 across
- 8 a small circle of DNA in a bacterium
- 9 can be used to carry genes into human lung tissue
- 10 an organism made by combining cells from two different species
- 11 these are growth chemicals needed to make plants grow well in culture media
- 12 a technique for transferring genes as a means of treating disease
- 14 a disease condition that can be helped by genetically engineered factor 8
- 16 with 18 across – the technique of using plant fragments to produce many identical plants
- 17 genetically modified bacteria are grown in large quantities inside this type of vessel
- 18 see 16 across
- 20 an important engineered protein used to treat diabetes
- 23 an enzyme needed to remove the cellulose cell wall from a plant cell
- 24 biological catalysts needed to 'cut and stick' pieces of DNA
- 26 the method of transferring useful genes
- 27 a group of genetically identical organisms, produced by mitosis
- 28 a plant cell with the cell wall removed

Down:

- 1 a method of 'concentrating' useful characteristics by carefully controlling sexual reproduction
- 2 the hereditary chemical (abbreviation)
- 3 the tip of a plant that can be broken up for use in tissue culture
- 4 a mammalian product used to deliver useful proteins
- 6 microbes that can be engineered to produce useful materials for humans
- 13 a product that is coded for by a gene
- 15 a technique for studying the cells of a developing fetus
- 19 the control centre of the cell
- 21 must be controlled to limit water loss by transplanted 'cloned' plants
- 22 a section of DNA that controls the production of a single protein
- 25 the world's most famous sheep

129

Revision: Analysing command words

Exam success: knowing what to do

Candidates taking an examination in IGCSE Biology are given instructions about what the examiner expects from them. These instructions are given in the introduction to each question, or to each part of a multi-part question. These instructions, which tell the candidate what to do, are given as **command words**.

For example, a candidate might be asked to **define** a biological term, to **describe** a biological process, or to **calculate** a numerical value. Define, describe, and calculate are command words. To be successful in an examination, a candidate must understand what each of the command words means.

Command words – What answer do you expect?

- Command words may require either concise answers or extended answers.
- Command words may require either recall or making logical connections between pieces of information.
- Some command words require only single word or single figure answers.

This list of the command words is taken from the IGCSE syllabus, published by CIE. The examination board stresses that these words are best understood when they are seen in an actual question, and they also point out that some other command words may be used. Even so, this list contains the most commonly used words and what they mean. In other words, these command words tell you what the examiner wants you to do when trying to answer a particular question.

Name: the answer is usually a technical term (diffusion, for example, or mitochondrion) consisting of no more than a few words. **State** is a very similar command word, although the answer may be longer, as a phrase or sentence. **Name** and **state** don't need anything added, i.e. *there is no need for an explanation*. Adding an explanation will take up time and probably won't gain any more marks!

> a. State **three** normal functions of a root.
>
> 1. *To anchor the plant in the soil*
> 2. *To absorb water from the soil*
> 3. *To absorb mineral ions from the soil*
>
> [3]

Define: the answer is a formal definition of a particular term. The answer is usually 'what is it' – for example, define the term active transport means 'what is active transport'.

Look carefully at how many marks are offered. It is often a good idea to add an example to a definition – in this way the examiner can be sure that you know what you are trying to define.

What do you understand by or **what is meant by** are commands that also ask for a definition, but again the marks offered suggest that you should add some relevant comment on the importance of the terms concerned.

List: you need to write down a number of points, usually of only one word, with no need for explanation. For example, you might be asked to list four characteristics of living organisms.

Revision

Analysing command words

Describe: your answer should simply say what is happening in a situation shown in the question, e.g. the number of germinating seeds increased to 55. There is no need for an explanation.

No call for explanation or comment ⟶

(c) ⟨Describe⟩ how water reaches a leaf and enters a palisade cell.

Evaporation/transpiration from leaf surface

water pulled up xylem to replace 'losses'

water moves by osmosis to palisade cell

water crosses palisade cell membrane by osmosis

[3]

Make **three** points for the **three** marks on offer ⟶

(d) ⟨Describe⟩ how sugar produced in a palisade cell reaches the roots.

Conversion to sucrose/transported in phloem/

unloaded in roots by diffusion or active transport

[2]

Explain: the answer will be in extended prose, i.e. in the form of complete sentences. You will need to use your knowledge and understanding of biological topics to write more about a statement that has been made in the question, or earlier in your answer.

The command word **explain** is often linked with **describe** or **state**, so that the examiner asks you to **describe and explain** or **state and explain**. This means that there are two parts to be answered – they can often be joined by the word because. For example 'the number of germinating seeds increased to 55 (describe) *because* the temperature has increased and germination is controlled by enzymes which are sensitive to temperature (explain)'.

Many answers fail to gain full marks because they do not obey both commands – they often **describe** but do not **explain**, so only gain half of the available marks.

---

c. There is concern that pollution of the environment may change the breeding grounds of the Adelie penguin.

⟨State and explain⟩ the effect this might have on the populations of the Leopard seal and the Ross seal.

Leopard seal *Population could fall because there would be less food for them (so they would breed less successfully)*

Ross seal *Population could rise because there would be fewer leopard seals to act as predators*

two pieces of information needed – can you link them with the word 'because'?

[4]

131

Revision — **Analysing command words**

Suggest: this command word has two possible meanings. In the first, you will need to use your biological knowledge and understanding to explain something that is new to you. You might use the principle of enzyme action (which you know about) to explain an industrial process (which might be new to you). Suggest also has a second meaning – it implies that that there may be more than one possible answer to a question. For example, there might be a number of different factors affecting the action of a digestive enzyme.

Questions which involve data response (like looking at a table of results, for example) and problem solving (like comparing three different environmental situations, for example) often begin with the command word **suggest**.

Calculate: a numerical answer is expected, usually obtained from data given in the question. Remember:

- show your working (there may be marks given for the correct method even if you get the wrong answer)
- give your answer to the correct number of significant figures, usually two or three
- give the correct units (if needed – sometimes these will already be given in the space left for your answer).

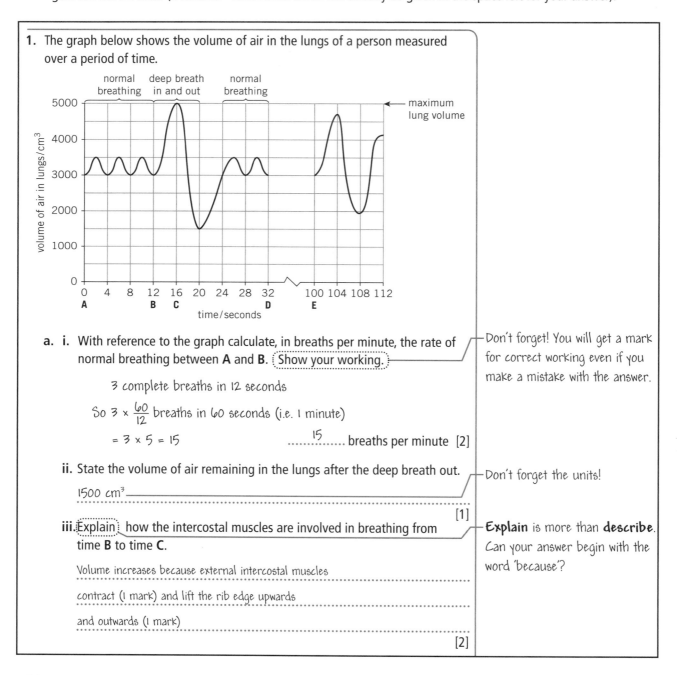

1. The graph below shows the volume of air in the lungs of a person measured over a period of time.

   a. i. With reference to the graph calculate, in breaths per minute, the rate of normal breathing between **A** and **B**. Show your working.

   3 complete breaths in 12 seconds

   So $3 \times \frac{60}{12}$ breaths in 60 seconds (i.e. 1 minute)

   $= 3 \times 5 = 15$

   ....15............ breaths per minute [2]

   — Don't forget! You will get a mark for correct working even if you make a mistake with the answer.

   ii. State the volume of air remaining in the lungs after the deep breath out.

   1500 cm³

   — Don't forget the units!

   [1]

   iii. **Explain** how the intercostal muscles are involved in breathing from time **B** to time **C**.

   Volume increases because external intercostal muscles contract (1 mark) and lift the rib edge upwards and outwards (1 mark)

   — **Explain** is more than **describe**. Can your answer begin with the word 'because'?

   [2]

Revision — Analysing command words

Other terms which require numerical answers are **find**, **measure**, and **determine**.

Find is a general term, and can mean calculate, measure, or determine.

Measure implies that the answer can be obtained by a direct measurement, e.g. using a ruler to measure the length of a structure on a diagram.

Determine means that the quantity cannot be measured directly, but has to be obtained by calculation or from a graph. For example, the size of a cell from a scale included in the diagram of the cell.

The graph below shows the heart rate and the cardiac output. The cardiac output is the volume of blood pumped out of the heart each minute.

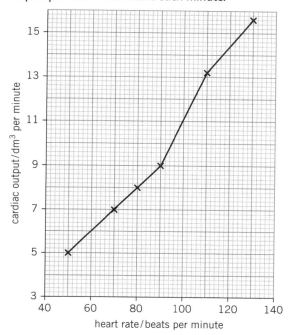

a. i. What is the cardiac output at a heart rate of 100 beats per minute? — *read directly (and accurately!) from the graph*

11 dm³ per minute

[1] *don't forget the units!*

ii. **Determine** the increase in cardiac output when the heart rate increases from 70 to 90 beats per minute.
  7   9

(9 − 7) = 2 .... dm³ per minute   [1]

means the same as 'calculate' (or 'work out')

iii. Determine the increase in cardiac output when the heart rate increases from 100 to 120 beats per minute.
  11   14.4

(14.4 − 11) = 3.4 .... dm³ per minute   [1]

Project ideas

24.1 Modelling neurones

ACTIVITY: MAKING MODEL NEURONES

You will need:

- Lengths (about 20–25 cm) of multi-strand single core wire, with plastic insulation. Ideally use wire with white or yellow insulation
- Small beads, with a hole for threading
- Wire strippers

Method:

1. **Motor neurone**

   - Strip 10 mm of plastic from the end of the first length of wire.
   - Push the exposed inner core through the thread hole in the bead. Push the bead up close to the plastic insulation, and then spread out the protruding strands from the inner core.
   - Strip 5 mm of plastic from the other end of the wire. Spread out the strands of the core wire that you have exposed.

2. **Sensory neurone**

   - Strip 25 mm of plastic insulation from one end of the wire. Thread on a bead as above, but only spread out the final 2 or 3 mm of inner core strands.
   - Strip 5 mm of plastic from the other end of the wire. Spread out the strands of the core wire that you have exposed.

3. Draw your two models. Use a library book or the internet to find diagrams of sensory and motor neurones so that you can label your models.

Project ideas

24.2 Modelling the spinal cord

ACTIVITY: A MODEL OF THE SPINAL CORD

You will need:

- Plasticine
- Single core, plastic insulated wire
- White beads
- Wire strippers

Method:

1. Make a Plasticine model of a cross section of a spinal cord. Use a different colour for the central grey matter.

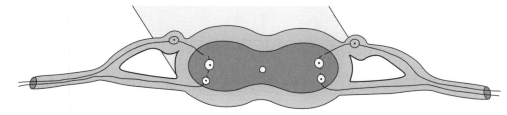

2. Make a sensory and motor neurone as described in unit 24.1.

3. Make an intermediate (connector) neurone. This only needs to be about 10 mm long, and doesn't have any insulation.

4. Arrange the neurones in the correct sequence on the model spinal cord. This now represents a single reflex arc.

Alternative method: if no Plasticine is available, you can draw out a cross-section of the spinal cord, and lay your three neurones on it.

Practical biology — 25.1 Variables

1. A student carrying out food tests for biological molecules had seen on the internet that **cow's milk contained more sugar than soya milk**. He suggested that the group should test this hypothesis.

   a. Describe how you would carry out the test for a simple sugar.

   ....

   ....

   .... [3]

   b. The test would only be valuable if it is **quantitative**. State the meaning of the term quantitative.

   .... [1]

   c. The teacher suggested that the students should fill in this table, to make sure that they had a plan that would provide them with valid data. Complete the table.

Which variable would you change in the experiment? This is the **independent variable**.	
Which three variables would you keep the same? These are the **fixed variables**.	
Which variable would you measure to test the hypothesis? This is the **dependent variable**.	
How would you measure this variable?	
Suggest a control for the experiment.	
Suggest why you would repeat the experiment.	

[8]

Practical biology

25.2 Making a model of DNA

Although it took Watson and Crick many years of work, you can create a DNA model quite easily! You will be using some of the same information that Watson and Crick had available to them.

- DNA is made up of simple subunits called **nucleotides**.
- There are only four different nucleotides found in DNA – **adenine** (A), **guanine** (G), **cytosine** (C), and **thymine** (T). There is always the same amount of adenine as thymine, and the same amount of guanine as cytosine.
- There are **base pairing** rules: A always pairs with T, and G always pairs with C.

Alright, you are ready to go!

1. Cut out the block of nucleotides below, and separate each nucleotide from the others.

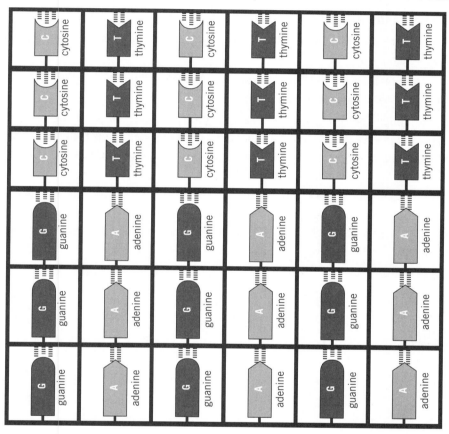

2. Place one each of the nucleotides in the spaces below. Arrange them so that the 'bonds' hold the nucleotides in the correct base pairs.

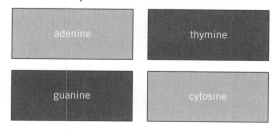

   a. Use the diagram to explain why A always pairs with T, and G always pairs with C.
   b. What do you notice about the two 'letters' in each of the pairs?

137

Practical biology   **25.2 Making a model of DNA**

3. Now create one strand of the DNA molecule. Do this by placing the correct cut-out nucleotide in the labelled spaces of the outline below.

4. Build up the 'matching' strand of DNA by using the base pairing rule (scientists use the word 'complementary' for 'matching'). Remember that the nucleotides in the complementary strand will be upside down.

CONGRATULATIONS! You have now produced a model of a short chain of DNA. If you are working as part of a large class group, perhaps some people in the class could build slightly different sequences, e.g.

AGGGCTTAAT   CCAGGCCTTA   TAGCCAGTAG

GGCATTGCCC   CGCGATATTG   AGCCCTTAGG

If possible, you should now make a copy of your piece of DNA. If your teacher has given you an outline for the DNA model, you can stick the nucleotide 'letters' into their correct positions.

Now you could try something really tricky! In 'real' DNA, the two strands are coiled around each other to make a **double helix**. The twisting takes place once every ten pairs, so....

> You can now twist your piece of DNA so that there is just one turn in it. You should be able to clearly see the top base pair and the bottom base pair. Now you have a piece of **double helix**.

> A gene is a section of DNA that codes for one protein in a cell. Even a small protein needs quite a long gene. If you join up your piece of DNA with pieces from your classmates, you might get a good idea of a single gene. For example, about 30 pieces joined together in a long double helix would give enough coded information to make a simple protein like **insulin** (the protein hormone which controls our blood sugar level).

A human cell contains more than 10 000 genes in its nucleus. You need so much DNA to make up these genes that each of your cells (which you can't even see without a microscope!) contains about 2 metres of DNA. The DNA has to coil up like balls of wool to fit in! A mature human has so many cells that there is enough DNA in each of us to stretch to the Moon... and back... and to the Moon again!

adenine	
guanine	
adenine	
cytosine	
guanine	
cytosine	
thymine	
thymine	
adenine	
guanine	

Practical biology
25.3 Drawing skills: the structure of flowers

A flower is made up of a set of modified leaves. The leaves are arranged in rings which produce the gametes, protect them, and ensure that fertilisation takes place. The rings of leaves are attached to the end of the flower stalk, the receptacle.

Examination of a typical flower

If a typical flower (a buttercup, for example) is examined, it is possible to see the four rings of specialised leaves. It may be necessary to use a hand lens to do this easily.

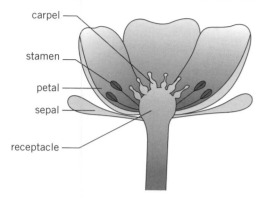

1. Arrange the four rings (carpel, sepal, stamen, and petal) from outside to inside.

   Use a scalpel (take care!) and a pair of forceps to remove a representative part of each ring of leaves. Examine the structure under a hand lens or binocular microscope.

   The individual structures will look something like those shown in this diagram.

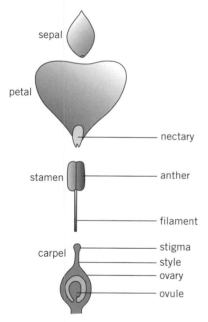

How many of each of the structures can you find in your flower?

Variations in flower structure

Flowers of different species may differ in:

- the number of each of the different components
- the arrangement of the different parts – especially how fused (stuck together) they are to form tubes or platforms.

Drawing a flower

With so much variation in flower structure, scientists have developed standard ways of describing flower structure. The most common method is to draw a **half flower**.

Use a razor blade or scalpel to cut through the flower, down the line of its stalk. Hold the flower in forceps if necessary, and take care with the sharp blades.

Practical biology — 25.4 Germination

1. You are asked to design an experiment to study the effect of temperature on the germination success rate of pea seeds. The teacher suggests a temperature range of 5–55 °C.

   a. State the independent, the dependent, and any fixed variables that you should include in your experimental design.

independent variable	dependent variable	fixed variables

   [5]

   b. Suggest how you would measure the value of the dependent variable. [1]

   c. Draw a table in which you could present your results.

   [3]

   d. Prepare a grid on which the results could be plotted. Draw a graph of the likely results of your investigation.

   [4]

   e. Explain the results which you have presented in the graph.

   [3]

Practical biology — 25.5 Transpiration experiment

1. The diagram shows a piece of apparatus that can be used to measure the rate of water uptake by a plant shoot.

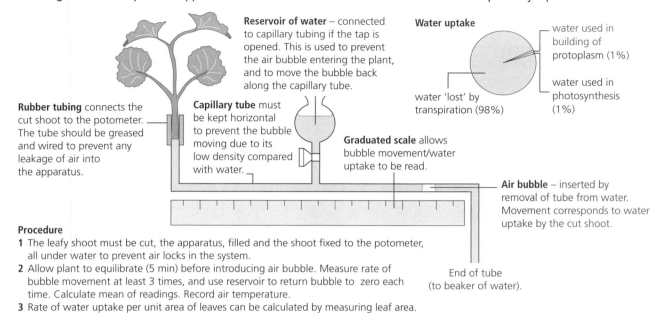

Procedure
1 The leafy shoot must be cut, the apparatus, filled and the shoot fixed to the potometer, all under water to prevent air locks in the system.
2 Allow plant to equilibrate (5 min) before introducing air bubble. Measure rate of bubble movement at least 3 times, and use reservoir to return bubble to zero each time. Calculate mean of readings. Record air temperature.
3 Rate of water uptake per unit area of leaves can be calculated by measuring leaf area.

Two sets of this apparatus were used at the same time. The two sets of apparatus had shoots taken from the same tree.

The experiment was carried out with four different sets of external conditions **A**, **B**, **C**, and **D**. The time for the air bubble to move 10 cm was measured and recorded in the table.

	external conditions	time for air bubble to move 10 cm / s	
		shoot 1	shoot 2
A	dry, still air at 15 °C	25	46
B	dry, still air at 25 °C	19	37
C	dry, moving air at 25 °C	16	32
D	humid, still air at 15 °C	58	78

a. State which shoot took up water most quickly under all conditions.

.... [1]

b. Suggest a difference between the shoots that could explain these results.

....

.... [1]

c. Explain why the results were different under condition **D** than from the condition in **A**.

....

....

.... [2]

141

Mathematics for biology

26.1 Measurement and magnification

The **size** of a structure or organism is measured in units of **length** (such as mm or m). When a diagram is made, or a photograph taken, it may not be easy to directly show the correct size – for example, when a structure is extremely small or very large.

The correct (or true) size of an organism can be calculated using a combination of actual measurement and a known magnification.

There are two simple relationships that should be understood:

Magnification = $\dfrac{\text{measured length}}{\text{actual length}}$

Actual (true) length = $\dfrac{\text{measured length}}{\text{magnification}}$

It is also important that candidates can use a **scale line** to work out magnification.

5 mm

This means that the line drawn represents 5 mm in actual length.

So, magnification = $\dfrac{\text{measured length of scale line}}{\text{actual length of scale line}}$

= $\dfrac{85}{5}$ = 17

Note that there are no units for magnification – it is a comparison of lengths. Be careful to make sure that the two lengths you are comparing are given in the same units.

1. State

   a. How many mm there are in 1 cm ....

   b. How many μm there are in 1 mm..

Core students (Papers 1 and 3) can use millimetres (mm) as units, but candidates taking extension papers (Papers 2 and 4) should also be confident with the use of **micrometres** (μ or μm).

2. The diagram shows a cell from the pancreas of a human. The cell was drawn using a light microscope.

   a. Identify the structures labelled **A**, **B**, and **C**. [3]

   b. Which of these structures would not be present in a red blood cell? [1]

   c. Name two additional structures that you would see in a palisade cell from a leaf. [2]

   d. Use the scale shown alongside the cell to calculate how much it has been magnified. Show your working. [3]

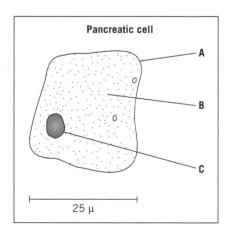

Pancreatic cell

25 μ

142

Mathematics for biology

26.2 Multiple births

1. Approximately 1% of all births are **multiple births**, i.e. more than one baby develops and is born at the same time.

   This table shows the frequency of multiple births in mothers in different age groups.

age group of mother (years)	frequency of multiple births (percentage of all live births)
younger than 20	0.6
20–24	0.7
25–29	1.0
30–34	1.2
35–39	1.5
40–44	1.2
older than 45	1.4

   a. Plot this information as a histogram. Use the grid provided below.

   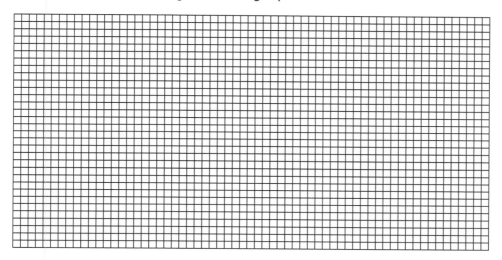

   b. Comment on the effect of the mother's age on the frequency of multiple births.

   ....

   ....

   .... [2]

143

Mathematics for biology

26.3 Enzyme experiments

1. The rate of activity of amylase is affected by temperature. The effect of temperature was investigated by using a set of six identical test tubes containing 5 cm³ of starch solution. Each test tube was placed in a water bath at a different temperature and then 1 cm³ of amylase solution was added to start the reaction.

   The time taken for the starch to disappear was measured. The results are recorded in this table.

temperature (°C)	20	25	30	35	40	45
time taken for starch to disappear (s)	600	315	210	175	200	420

   a. Use the grid provided to plot a graph of these results.

   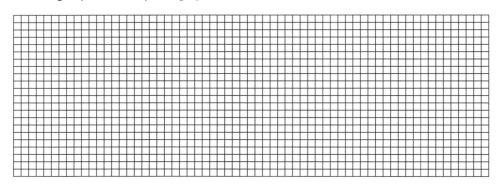

   [4]

   b. State the temperature at which the amylase works best.

   ....°C [1]

   c. Explain why it is important that the same volume of starch and amylase was present at the start of the experiment.

   ....

   ....

   .... [1]

   d. Chemical reactions usually get faster as temperature increases. Suggest a reason why the rate of amylase activity does not increase above 40°C.

   ....

   ....

   .... [2]

   e. Name one other factor that would affect the rate of amylase activity.

   .... [1]

Mathematics for biology

26.4 The control of photosynthesis

1. Scientists were able to grow red peppers and lettuces in large greenhouses. This table shows the effects on crop yields of adding extra carbon dioxide to the atmosphere inside the greenhouses.

crop	crop yield (kg)	
	normal air (0.04% CO_2)	enriched air (0.08% CO_2)
red peppers	0.42	0.63
lettuces	0.9	1.1

   a. Calculate the percentage increase in yield for red peppers when the air is enriched with carbon dioxide. Show your working.

   .... [2]

   b. There are a number of possibilities for increasing the carbon dioxide concentration in the greenhouse. Study this list, then choose the best method. Explain your decision.

   - Relying on the exhaled carbon dioxide from greenhouse workers.
   - Adding extra manure to the soil so that microbes can respire and release carbon dioxide.
   - Releasing carbon dioxide from cylinders of compressed gas.
   - Using paraffin heaters in the greenhouse, as burning paraffin releases carbon dioxide.

   Best method ....

   Explanation ....

   .... [2]

145

Mathematics for biology

26.5 Ingestion

1. The pH of the saliva of 100 students in a school was measured and compared with the number of teeth which had required fillings.

   The results are presented in this table.

pH of saliva	percentage of pupils in each pH range	average number of fillings per student in each pH range
6.7–6.9	20	8
7.0–7.2	60	5
7.3–7.5	20	3

   a. Suggest how the pH of the saliva could be measured.

   ....

   .... [2]

   b. State the hypothesis being tested in this investigation.

   .... [1]

   c. State and explain whether the results support the hypothesis.

   ....

   ....

   .... [2]

Exam-style questions for IGCSE

Multiple choice

Remember that there are two alternative multiple choice papers. Paper 1 contains questions which only assess material from the **core** syllabus; paper 2 contains questions that could contain material from either the **core** or the **supplement**.

Sample 'core' questions

1. Root hair cells are found on plant roots.

   Which of the following features would be present in a root hair cell but not in a cell lining the small intestine?

   A cytoplasm
   B chloroplasts
   C cell wall
   D cell membrane

2. Which animal is a spider?

   1. has legs .... go to 2

      has no legs .... go to 3

   2. has six legs .... organism **A**

      has eight legs .... organism **B**

   3. has a shell .... organism **C**

      has no shell .... organism **D**

3. Dietary fibre passes through several structures after leaving the stomach.

   In which order does the dietary fibre pass through these structures?

   A ileum – duodenum – rectum – colon

   B duodenum – ileum – rectum – colon

   C ileum – duodenum – colon – rectum

   D duodenum – ileum – colon – rectum

4. Which chemical elements are found in carbohydrates, fats, and proteins?

	carbohydrates	fats	proteins
A	carbon, hydrogen, and oxygen	carbon, hydrogen, and oxygen	carbon, hydrogen, oxygen, and nitrogen
B	carbon, hydrogen, and oxygen	carbon, hydrogen, oxygen, and nitrogen	carbon, hydrogen, and oxygen
C	carbon, hydrogen, oxygen, and nitrogen	carbon, hydrogen, and oxygen	carbon, hydrogen, and oxygen
D	carbon, hydrogen, oxygen, and nitrogen	carbon, hydrogen, and oxygen	carbon, hydrogen, oxygen, and nitrogen

Exam-style questions for IGCSE

5. The diagram shows a germinated bean seed with a horizontal radicle. This is placed on a slowly rotating disc and is left for three days.

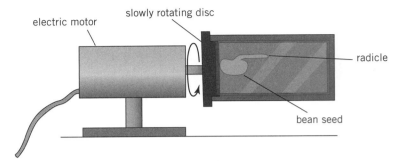

Which diagram shows the appearance of the radicle after three days?

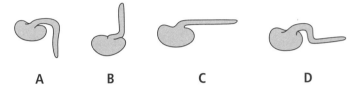

6. The diagram shows a section through the eye.

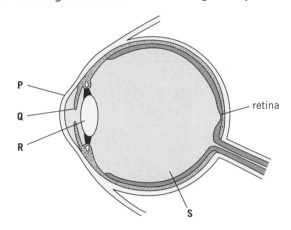

Which pair of structures focus light rays onto the retina?

A P and Q   B P and R   C Q and R   D Q and S

Sample 'supplement' questions

7. The graph shows the rate of growth for a population of herbivores.

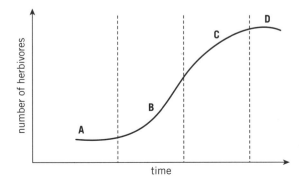

Which is the exponential (log) phase for the growth of the population?

8. Which graph shows the effect of temperature on the rate of photosynthesis?

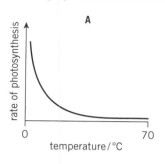
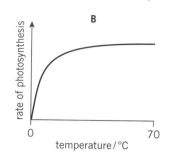
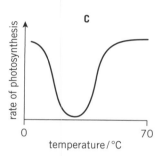
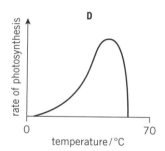

9. The diagram shows part of a transverse section of a leaf.

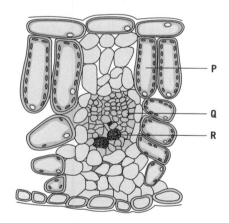

Which cells conduct water into the leaf and which cells conduct sugars out of the leaf?

	conduct water	conduct sugars
A	P	Q
B	Q	P
C	Q	R
D	R	Q

149

10. The diagram shows four places on a river. Water samples were taken at each place.

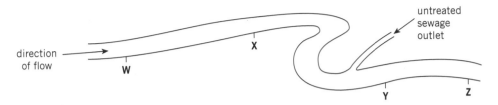

Which graph shows oxygen concentrations in the river?

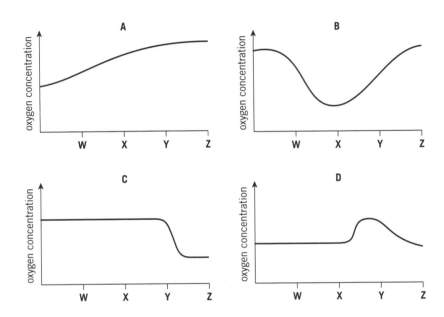

Exam-style questions for IGCSE

Longer written answers

Remember that there are also two alternative written papers. Paper 3 contains questions that only assess material from the **core** syllabus; paper 4 contains questions that could contain material from either the **core** or the **supplement**.

Sample 'core' questions

11. The diagram below shows the human alimentary canal.

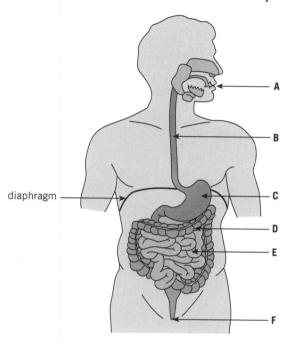

a. On the diagram, draw the liver in the correct place and approximately the right size. [2]

b. State the name of the process that moves food along the gut.

.... [1]

c. Use the diagram to complete the following table.

Each letter may be used once, more than once, or not at all.

description of site	letter
where saliva is released	
where food is chewed	
where hydrochloric acid is produced	
where soluble foods are absorbed	
where faeces are egested	

[5]

Exam-style questions | Exam-style questions for IGCSE

12. The diagram below shows how carbon may be recycled.

    The letters represent processes that occur in the cycle.

    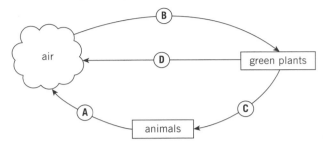

    a. i. Use words from the following list to identify the processes **A**, **B**, **C**, and **D**.

    feeding **photosynthesis** **respiration**

    The words may be used once, more than once, or not at all.

letter	process
A	
B	
C	
D	

    [4]

    ii. Name and describe **one** other process which is part of a complete carbon cycle.

    ....

    ....

    ....

    .... [3]

b. The diagram below shows a compost heap and the materials used to make it.

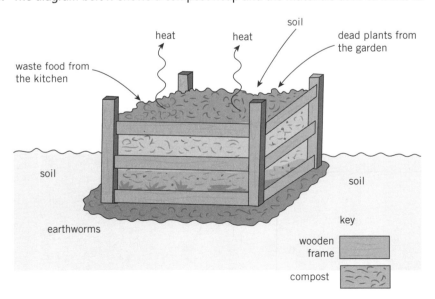

i. Explain why there is a difference in the temperature between the compost heap and the air surrounding it.

....

....

.... [2]

ii. Suggest one reason why a gardener uses an open frame to support the compost heap, rather than close the sides of the frame.

.... [1]

13. The diagram below shows a section through an air sac and a surrounding blood capillary.

   a. i. List three features shown in the diagram that make an air sac an efficient site for the exchange of gases.

   1. ....

   2. ....

   3. .... [3]

   ii. Suggest why the walls of the air sac contain elastic fibres.

   .... [1]

   b. Use the bar graph to identify **two** differences between the composition of the blood at **A** compared with the composition of the blood at **B**.

   A and **B** are shown in the diagram for part **a**.

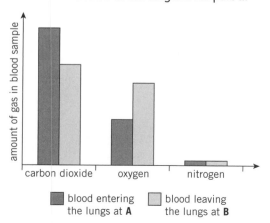

1. ....
   ....

2. ....
   .... [2]

c. Suggest **three** ways in which smoking can reduce the efficiency of the lungs.

   1. ....
      ....

   2. ....
      ....

   3. ....
      .... [3]

14. Phenylketonuria is a disease affecting the central nervous system of humans.

    This disease is caused by a single gene mutation.

    a. i. Define the term *mutation*.

       ....
       .... [1]

       ii. State the name of **one** other disease affecting humans which is the result of a gene mutation.

       .... [1]

    b. The diagram shows part of a family tree showing phenylketonuria.

    key:
    ■ affected male   □ normal male
    ● affected female ○ normal female

154

i. Phenyketonuria is caused by a recessive allele, **p**.

Define the term *recessive*. ....

.... [2]

ii. State the phenotype of individual 8. .... [1]

iii. State the possible genotypes of: individual 4 ....

individual 6. .... [2]

iv. Individual 10 marries a man who is heterozygous for this condition.

Use a Punnett square to find the probability that their first child will have phenylketonuria.

Probability .... [4]

Glossary

Absorption: Movement of small food molecules and ions through the wall of the intestine into the bloodstream.

Adaptive feature: An inherited feature that helps an organism to survive and reproduce in its environment.

Adaptive feature: The inherited functional features of an organism that increase its fitness.

Active immunity: Defence against a pathogen by antibody production in the body.

Active transport: The movement of particles from a region of lower concentration to one of higher concentration, across a membrane, using energy from respiration.

Aerobic respiration: The chemical reactions in cells that use oxygen to break down nutrient molecules to release energy.

Allele: A version of a gene.

Anaerobic respiration: The chemical reactions in cells that break down nutrient molecules to release energy without using oxygen.

Asexual reproduction: A process resulting in the production of genetically identical offspring from one parent.

Assimilation: The use of digested food molecules by cells.

Binomial system: Naming species in a system in which the scientific name of an organism is made up of two parts showing the genus and species.

Carnivore: An animal that gets its energy by eating other animals.

Catalyst: A substance that increases the rate of a chemical reaction and is not changed by the reaction.

Chemical digestion: Breakdown of large, insoluble food molecules into small, soluble molecules.

Chromosome: A thread-like structure of DNA, carrying genetic information in the form of genes.

Community: All of the populations of different species in an ecosystem.

Consumer: An organism that gets its energy by feeding on other organisms.

Cross-pollination: Transfer of pollen grains from the anther of a flower to the stigma of a flower on a different plant of the same species.

Deamination: The removal of the nitrogen-containing part of amino acids to form urea.

Decomposer: An organism that gets its energy from dead or waste organic material.

Diffusion: The net movement of particles from a region of higher concentration to one of lower concentration down a concentration gradient, as a result of their random movement.

Diploid nucleus: A nucleus containing two sets of chromosomes, e.g. in body cells.

Dominant: An allele that is expressed if it is present.

Drug: Any substance taken into the body that modifies or affects chemical reactions in the body.

Ecosystem: A unit containing the community of organisms and their environment, interacting together, e.g. a decomposing log, or a lake.

Egestion: The passing out of food that has not been digested or absorbed, as faeces through the anus.

Enzyme: A protein that functions as a biological catalyst.

Excretion: Removal of toxic materials and substances in excess of requirements.

Fertilisation: The fusion of gamete nuclei.

Fitness: The probability of an organism surviving and reproducing in the environment in which it is found.

Food chain: The transfer of energy from one organism to the next, beginning with a producer.

Food web: A network of interconnected food chains.

Gene: A length of DNA that codes for a protein.

Gene mutation: A change in the base sequence of DNA.

Genetic engineering: Changing the genetic material of an organism by removing, changing, or inserting individual genes.

Genotype: The genetic make-up of an organism in terms of the alleles present.

Gravitropism: A response in which parts of a plant grow towards or away from gravity.

Growth: A permanent increase in size and dry mass by an increase in cell size and/or cell number.

Haploid nucleus: A nucleus containing a single set of unpaired chromosomes, e.g. in gametes.

Herbivore: An animal that gets its energy by eating plants.

Heterozygous: Having two different alleles of a particular gene.

Homeostasis: The maintenance of a constant internal environment.

Homozygous: Having two identical alleles of a particular gene.

Hormone: A chemical substance, produced by a gland and carried by the blood, which alters the activity of one or more specific target organs.

Ingestion: The taking in of substances into the body through the mouth.

Inheritance: The transmission of genetic information from generation to generation.

Glossary

Limiting factor: Something present in the environment in such short supply that it restricts life processes.

Mechanical digestion: Breakdown of food into smaller pieces without chemical change to the food molecules.

Meiosis: Nuclear division giving rise to cells that are genetically different.

Mitosis: Nuclear division giving rise to genetically identical cells.

Movement: An action causing the change in position of an organism or part of an organism.

Mutation: Genetic change.

Nutrition: The taking in of materials for energy, growth, and development.

Organ: A structure made up of a group of tissues, working together to perform a specific function.

Organ system: A group of organs with related functions.

Osmosis: The net movement of water molecules from a region of higher water potential to a region of lower water potential, through a partially permeable membrane.

Passive immunity: Short-term defence against a pathogen by antibodies acquired from another individual, e.g. mother to infant.

Pathogen: A disease-causing organism.

Phenotype: The observable features of an organism.

Photosynthesis: The process by which plants manufacture carbohydrates from raw materials using energy from light.

Phototropism: A response in which parts of a plant grow towards or away from the direction from which light is coming.

Pollination: Transfer of pollen grains from anther to stigma.

Population: A group of organisms of one species, living in the same area, at the same time.

Process of adaptation: The process, resulting from natural selection, by which populations become more suited to their environment over many generations.

Producer: An organism that makes its own organic nutrients, usually using energy from sunlight, through photosynthesis.

Recessive: An allele that is only expressed when there is no dominant allele of the gene present.

Reproduction: The processes that make more of the same kind of organism.

Respiration: The chemical reactions in the cell that break down nutrient molecules to release energy for metabolism.

Self-pollination: The transfer of pollen grains from the anther of a flower to the stigma of the same flower or different flower on the same plant.

Sense organ: Groups of receptor cells responding to specific stimuli: light, sound, touch, temperature, and chemicals.

Sensitivity: The ability to detect and respond to changes in the environment.

Sex-linked characteristic: A characteristic in which the gene responsible is located on a sex chromosome.

Sexual reproduction: A process involving the fusion of the nuclei of two gametes (sex cells) to form a zygote and the production of offspring that are genetically different from each other.

Sexually transmitted infection: An infection that is transmitted via body fluids through sexual contact.

Species: A group of organisms that can reproduce to produce fertile offspring.

Sustainable development: Development providing for the needs of an increasing human population without harming the environment.

Sustainable resource: One which is produced as rapidly as it is removed from the environment so that it does not run out.

Synapse: A junction between two neurones.

Tissue: A group of cells with similar structures, working together to perform a shared function.

Translocation: The movement of sucrose and amino acids in the phloem.

Transmissible disease: A disease in which the pathogen can be passed from one host to another.

Transpiration: Loss of water by plant leaves by evaporation and diffusion.

Trophic level: The position of an organism in a food chain, food web, pyramid of numbers, or pyramid of biomass.

Variation: Differences between individuals of the same species.

Answers

Unit 1.1
LL true
1. A: sensitivity, B: reproduction, C: excretion, D: respiration
2. C
3. kingdom – phylum – class – order – family – genus – species

Unit 1.2
LL mammalia; genus; species
1. a. i. *Parus caeruleus* and *Parus major* they belong to the same genus
    b. i.
        E. rubecula   X X X
        P. caeruleus X ✓ ✓
        P. major   X ✓ ✓
        T. merula   X X X
        ii. Pale area only below eye *Parus major*
            Pale areas above and below eye *Parus caeruleus*

Unit 1.3
LL Literal meaning is 'before nucleus', and refers to the absence of a nuclear membrane in these organisms
1. a. i. chloroplast   ii. nutrition (photosynthesis)
    b. i. excretion   ii. movement
    c. respiration

Unit 1.4
LL vertebrates; classes; mammals; reptiles; fish
1. a. An animal with a backbone
    b. Scales; no; no; yes; fur – yes

Unit 1.5
LL 1. false 2. false 3. true 4. false
1. a. Ant: three/yes; earthworm: none/no; centipede: many/yes; mite: four/no
    b. Ant – insect; earthworm – annelid; centipede – myriapod; mite – arachnid

Unit 1.6
LL stem; leaves; cellulose; ferns
1. a. Stem – hold leaves in best position; root – absorb water and mineral ions; leaves – trap light energy for photosynthesis; flowers – may be attractive to pollinating insects or birds; fruit – usually help dispersal of seed, a reproductive structure
    b. Chlorophyll   autotrophic   photosynthesis   cellulose   algae   ferns   angiosperms   monocotyledons   dicotyledons

Unit 1.7
LL true
1. A: *Thynnus*   B: *Exocoetus*   C: *Paracirrhites*   D: *Scyliorhinus*

Unit 2.1
LL Plant cells have a rigid cellulose cell wall, which fixes their shape. Animal cells have only a flexible cell surface membrane.
1. a. A: nucleus   B: chloroplast
    b. i. show a granule (mitochondrion) in the cytoplasm
        ii. mitochondrion
    c. Measured diameter = 30 mm = 30 000 μ.
        Actual diameter = 25 μ.
        Magnification = $\frac{30\,000}{25}$ = 1200
    d. partially/differentially permeable membrane

Unit 2.2
LL nucleus – site of DNA; ribosome – protein synthesis; mitochondrion – respiration; cell membrane – active transport
1. a. phloem – transport; stamens – reproductive.
    b. i. Could label nucleus/cell membrane/cytoplasm
        ii. chloroplast
        iii. cellulose cell wall/permanent vacuole
    c. i. mitochondrion
        ii. ribosomes (accept rough endoplasmic reticulum)

Unit 2.3
LL red blood cell – oxygen transport; neurone – impulse conduction; phloem sieve tube – sucrose transport; root hair cell – ion uptake
1. a. From top line, reading right to left:
        Cell C, cell B, cell F, cell I, cell E, cell A, cell H, cell D, cell G
    b. leaf

Unit 2.4
LL D: tissues – systems - organs
1. a. heart: circulatory   stomach: digestive   bladder: urinary/excretory
    b. Phloem – transport; stamens – reproductive

Unit 3.1
LL It is partially correct – diffusion does NOT need the presence of a partially permeable membrane
1. Diffusion   down   gas   random   equilibrium
    Osmosis   diffusion   potential   partially permeable

Unit 3.2
LL diffusion; water; partially permeable membrane; flaccid; turgid
1. a. i. +8, +4, –1, –6, –10
        ii. graph plotted
        iii. 0.45
        iv. Represents an equilibrium so shows water potential of cell cytoplasm
    b. osmosis

Unit 3.3
LL Flaccid; osmosis; cellulose; haemolysis
1. a. Rigid due to hydrostatic pressure against inside of cellulose cell wall
    b. Offers support – diagram should show protoplast pushing against inside of cellulose cell wall
    c. i. A
        ii. The potato chip would straighten/become rigid as water would enter (down the water potential gradient) by osmosis

Unit 3.4
LL C and D are true
1. a. i. B
        ii. A
    b. D
    c. Movement is against a concentration gradient
    d. 1. Mineral ion, e.g. magnesium ion into root hair
        2. Glucose/amino acids from gut contents across villi into bloodstream

Unit 4.1
LL True
1. a. glucose   cellulose   sucrose   soluble.
    b. fatty acids   glycerol   insoluble.
    c. haemoglobin   amino acids   soluble.
    d. DNA
2. Protein: + + + – – +;
    Fat: + + + – – –;
    Carbohydrate: + + + – – –;
    Nucleic acid: + + + – + +

Unit 4.2
LL Protein – enzymes are examples; fat – are insoluble in water; carbohydrate – contain approximately twice as many hydrogen atoms as oxygen atoms; vitamin C – can oxidise DCPIP
1. a. i. higher protein content
        ii. Earlier step in food chain, so more can be produced per unit area. Beef cattle must live longer before they can be killed for meat.
    b. i. sample Z
        ii. as a reagent blank
        iii. sample X

Answers

Unit 4.3
LL deoxyribonucleic acid; Watson; Crick; Franklin
1. a. From the top of the table: TRUE; FALSE; TRUE; TRUE; TRUE; TRUE; FALSE; TRUE; TRUE; TRUE
   b. Obviously there will be different arguments in different groups.

Unit 5.1
LL 1 – true; 2 – false; 3 – false
1. Protein – the type of molecule that makes up an enzyme; substrate – a molecule that reacts in an enzyme-catalysed reaction; product – the molecule made in an enzyme-catalysed reaction; active site – the part of the enzyme where substrate molecules can bind; denaturation – a change in shape of an enzyme so that its active site cannot bind to the substrate; optimum – the ideal value of a factor, such as temperature, for an enzyme to work
2. Lipase; cuts out useful genes from chromosomes; removes milk sugar from milk; protease; cellulase; amylase

Unit 5.2
LL The loss of the three dimensional structure of an enzyme (particularly its active site), usually caused by extremes of temperature or pH
1. a. Check correct column headings e.g. temperature (°C) and time for clotting (minutes)
   b. Graph plotted – temperature on x axis
   c. 45 °C
   d. i. pH
      ii. use buffer solutions

Unit 5.3
LL optimum; stomach; duodenum/ileum/small intestine
1. a. i. amino acids/peptides
      ii. 54 mg
   b. i. 7.0
      ii. 2.4
   c. Hydrochloric acid is added from cells lining the glands of the stomach
   d. i. lipase
      ii. fatty acids and glycerol
      iii. Bile, by breaking it down to fat droplets (emulsification) increases surface area for action of lipase

Unit 6.1
LL water; glucose
1. photosynthesis   light/solar   chloroplasts/chlorophyll carbon dioxide   water   starch   oxygen   stomata
2. a. carbon dioxide
   b. oxygen
   c. water
   d. nitrate
   e. magnesium

Unit 6.2
LL starch; iodine; brown; blue-black
1. a. to destarch it (remove any starch already present)
   b. magnesium
   c. chlorophyll (necessary to absorb light energy) or light energy (needed to split water to begin production of carbohydrate)

Unit 6.3
LL 1 – false; 2 – true
1. a. starch
   b./ c. Starch gives a blue-black colour with iodine solution (normally brown in absence of starch)
   d. To supply energy by respiration/to produce cellulose for cell wall structure

Unit 6.4
LL C: Oxygen concentration
1. a. 3.5 arbitrary units
   b. Light intensity is the limiting factor up to 6 arbitrary units. At this point, some other factor (temperature, for example) is at a value which is preventing further photosynthesis.

Unit 6.5
LL B: raise carbon dioxide concentration
1. a. Light intensity 10 arbitrary units, carbon dioxide concentration 0.15% at 30°C. Make sure that you give all of the units in your answer!
   b. Even at the same light intensity and temperature, increasing the carbon dioxide concentration from 0.04% to 0.15% increased the rate of photosynthesis by as much as 375%.
   c. Carbon dioxide concentration is the limiting factor under those sets of conditions

Unit 6.6
LL B: stomata
1. a. A – upper epidermis   B – palisade mesophyll   C – spongy mesophyll   D – guard cell   E – xylem vessel
   b. B
   c. E
   d. sucrose
   e. i. 0.9 mm
      ii. 0.25 mm

Unit 6.7
LL Magnesium; nitrate; active transport
1. a. i. To allow measurement of the maximum (complete solution) and minimum (water) rate of growth of the wheat seedlings.
      ii. Air contains oxygen and oxygen is needed for aerobic respiration. This releases energy for active uptake of mineral ions by the root hair cells.
      iii. Magnesium is required to produce chlorophyll. No chlorophyll means very limited photosynthesis so very poor growth.
   b. nitrogen   phosphorus   potassium

Unit 7.1
LL C: carbohydrates
1. a. 70% = 55% + 15% (each segment represents 5%)
   b. From the top of the list: T; F; T; F; T; F; T; F; F; T; F; F; T; F

Unit 7.2
LL B: citrus fruits
1. a. carbohydrate and fat/lipid
   b. Carbohydrate, since pasta is largely carbohydrate and skimmed milk has no fat, so lower energy content

Unit 7.3
LL Obesity; diabetes
1. a. y axis energy requirement (kj/day × 1000), x axis type of person
   B. age (teenage boy needs more energy than eight year old); gender – male IT worker needs more energy than female IT worker; activity – manual worker needs more energy than IT worker
   c. Female is lighter than a male, so less mass to move about

Unit 7.4
LL rickets
1. a. i. male
      ii. female, aged 20–24
   b. i. as fat
      ii. diabetes/arthritis/heart disease
   c. Provision of an unbalanced diet
   d. i. protein required for growth/production of enzymes/haemoglobin
      ii. lethargy/poor rate of growth/inactivity

Unit 7.5
LL C: lipase
1. a. With iodine solution: straw-brown   blue-black   blue-black   blue-black
   With Benedict's reagent: orange-red   blue   blue   blue
   b. Mouth (from salivary glands), small intestine (from pancreas)
   c. This is the optimum temperature for amylase activity
   d. In the stomach

Answers

Unit 7.6
LL Molar – crushing; incisor – biting; canine – killing/piercing
1. a. i. A – dentine   B – cement   C – crown
        ii. blood vessels/nerves
    b. i. the enamel is so hard
        ii. nerves are inside the pulp cavity
    c. Vitamin C

Unit 7.7
LL Saliva; amylase; hydrogencarbonate; alkaline
1. a. From top of diagram: epiglottis   entrance to trachea   contracted circular muscle   bolus of food   contracted circular muscle relaxed circular muscle   stomach   pyloric sphincter
    b. Lubricate bolus/ moisten food/ add amylase/ add alkaline hydrogencarbonate

Unit 7.8
LL D: villi
1. a. villus
    b. small intestine/ileum
    c. i. M – Should be linked to lacteal (at centre of villus)
        ii. N – should be linked to surface epithelium (produces mucus from goblet cells)
    d. Large surface area increases rate of absorption, thin epithelium reduces distance for absorption, capillaries/lacteals remove absorbed products
    e. i. × 50
        ii. 1.5 mm

Unit 7.9
LL Colon; appendix; rectum; anus
1. a. i. colon
        ii. rectum
        iii. small extension of caecum on LH side of diagram
    b. i. cholera
        ii. diarrhoea – colon is unable to reabsorb water due to a salt imbalance
    c. Salt solution and glucose are taken as a solution. Salts readjust osmotic balance and glucose provides energy for active transport of the mineral ions.

Unit 8.1
LL A – sucrose; B – xylem; C – epidermis; D – phloem
1. a. A: phloem   B: xylem   C: epidermis
    b. i. B because this is xylem which transports water and water-soluble dyes such as eosin
        ii. Radioactive carbon dioxide is converted to radioactive carbohydrate by photosynthesis. The carbohydrate is transported, as sucrose, in the phloem. The phloem therefore shows up by fogging the photographic plate.

Unit 8.2
LL Support; photosynthesis; roots/root hairs; leaves/stomata
1. Osmosis   hairs   surface area   ions   magnesium   diffusion active transport   support   solvent   photosynthesis

Unit 8.3
LL False (water only)
1. a. Mean mass at start = 220.4 g. Mean mass after 24 h = 210.2 g
        Therefore mean loss of water = 220.4 – 210.2 = 10.2 g
    b. Mean volume of water at start = 100 cm^3 = 100 g
        Mean volume of water after 24 h = 88 cm^3 = 88 g
        Therefore mean uptake of water = 12 g
    c. Water uptake id driven by water loss from the stomata/leaves. Water lost 'pulls' a stream of water through the plant – this is replaced by water uptake at the roots.
    d. Some of the water absorbed by the roots is used for support or in photosynthesis, so the masses are not exactly the same.

Unit 8.4
LL aphids
1. a. A: xylem on top, phloem below   B: xylem on top, phloem below
    b. i. leaf – arrow towards stem; stem arrow at both ends of line; root – arrow down
        ii. glucose / sugars

Unit 9.1
LL Aorta – main artery in the body; renal artery – blood input to kidney; hepatic vein – blood output from liver; pulmonary artery – from heart to lungs
1. a. i. pulmonary vein
        ii. A has less oxygen/more carbon dioxide/is at higher pressure
    b. E has more urea/more sodium/is at higher pressure
    c. Should be guideline to vessel just below chamber C

Unit 9.2
LL 1 – true; 2 – false
1. a. i. left ventricle
        ii. thicker muscular walls
    b. Pacemaker …… 72
2. a. double
    b. Blood pressure falls as blood passes through the gills, and cannot be raised again before blood reaches body tissues
    c. kidneys
    d. Blood at low pressure can flow more easily through the wider veins
    e. Pulmonary arteries; tissue fluid/plasma

Unit 9.3
LL Increase in heart rate and increase in stroke volume
1. a. Change the depth of heartbeat (stroke volume), i.e. volume of blood pumped with each heartbeat
    b. i. X – cuspid valve, pressure in ventricle exceeds pressure in atrium so valve closes to prevent backflow
            Y – semilunar valve closes, as pressure in left ventricle falls below pressure in aorta, to prevent backflow of blood into heart
        ii. in the veins
        iii. Maintain one way flow of blood at lower pressure in the veins
    c. Blood flow to lungs is reduced so blood does not oxygenate well, and so blood (which shows up in the lips) is not as red as it normally would be

Unit 9.4
LL Pressure is lower in a vein, so no excess loss of blood
1. a. i. aorta
        ii. arterioles
    b. i. capillaries
        ii. large cross-sectional area/ low speed of travel of blood

Unit 9.5
LL Low vitamin D concentration
1. a. i. Exercise increases blood flow to heart muscles and skin, but reduces blood flow to gut and kidneys. Flow to brain does not change.
        ii. Blood delivers more oxygen and glucose for respiration needed to release the energy required for muscle contraction.
    b. i. Less oxygen is delivered to working heart muscle, so less energy is available and muscle stops working
        ii. obesity/high fat consumption/high salt consumption

Unit 9.6
LL 1 – false; 2 – true; 3 – true
1. a. i. 100 – 45 = 55%
        ii. ions /glucose/ amino acids/ hormones
        iii. red blood cell – transport of oxygen; phagocyte – engulfing invading microbes; lymphocyte – antibody production; platelet – part of clotting process
    b. Severe breathlessness/tiredness/inability to train hard
    c. They will have an increased number of red blood cells, so their blood will be able to supply an increased amount of oxygen to the muscles, thus improving their performance.

Answers

  d. Testosterone (or some other steroid) – used because it increases growth of muscle tissue; anabolic steroid / muscle building; stimulants such as amphetamine / more powerful muscle contractions

Unit 9.7
LL   B: clotting of blood
1. Pathogen – a disease-causing organism; transmissible disease – condition in which the pathogen can be passed from one host to another; antigen – a substance which triggers the immune response; skin and nasal hairs – external barriers to infection; active immunity – defence against a pathogen by antibody production in the body; passive immunity – short term defence by provision of antibodies from another individual; phagocyte – white blood cell that engulfs pathogens; lymphocyte – cell which produces antibodies
2. a. A chemical or cell which provokes the body's immune system to produce appropriate antibodies
   b. Response is faster/greater/lasts for longer

Unit 9.8
LL   1 – true; 2 – false; 3 – false
1. a. Blood returns to the heart twice for each complete circuit of the body
   b. i. Blood pressure falls, from 12 kPa to 4 kPa
      ii. from plasma to tissue fluid: glucose/amino acids/oxygen/ions such as iron and calcium; from tissue fluid to plasma: carbon dioxide/lactic acid/urea
   c. Blood pressure would fall (temporarily)
   d. Poor filtration of blood at kidneys so build up of toxic waste
   e. Large surface area/wall is only one cell thick

Unit 10.1
LL   D: a pathogen
1. a. High temperature/vomiting/diarrhoea
   b. Not washing hands when preparing food for child/not using clean utensils and crockery/feeding with infected food
   c. Always wash hands when preparing food for child/cook food thoroughly and at the correct temperature
2. Rickets/scurvy; cystic fibrosis/sickle cell anaemia; dementia/heart disease/cancer; lung cancer/CHD/diabetes type II

Unit 10.2
LL   An antigen is a substance which provokes activity of the immune system, and an antibody is a protein released by lymphocytes and capable of recognising an antigen
1. a. from left to right: lymphocyte   pathogen   antibody   antigen
   b. The antibodies may penetrate the pathogen and allow it to burst as water enters/phagocytes can be alerted by the presence of the antibody, and can then engulf and digest the pathogen
   c. vaccination

Unit 10.3
LL   B: vaccine
1. a. Defence against a pathogen by the production of antibodies in the body
   b. Antibodies are from another individual/the response does not last as long/no memory cells are produced.
      active   passive   passive   passive   active

Unit 10.4
LL   A: an antibiotic
1. a. Food source column: fish; milk products; fresh meat; vegetables. Number of reported outbreaks column: 20; 45; 5
   b. At refrigeration temperatures, bacteria which might be present in the food cannot multiply, so cannot produce toxins which cause food poisoning

Unit 11.1
LL   Bronchioles – bronchi – trachea
1. a. M: trachea   N: bronchus   O: bronchiole
   b. Label lines to intercostal (between the rib) muscles and the diaphragm (below lungs)
   c. growth/cell division/active transport/heat generation
   d. carbon dioxide/water vapour

Unit 11.2
LL   alveoli/air sacs; diffusion; carbon dioxide
1. a. Oxygen from alveolus to blood, and carbon dioxide from blood to alveolus.
   b. Large surface area/moist lining/very thin walls/closeness to capillary system
   c. pulmonary artery

Unit 11.3
LL   1 – false; 2 – false
1. a. lung
   b. Label a line drawn across the thorax below the lung
   c. external intercostal muscles
   d. Intercostal muscles relax so front of thorax falls. Diaphragm relaxes so bottom of the thorax rises. Lung volume falls, so air is pushed out (exhaled).

Unit 11.4
LL   B; 1600 cm³
1. a. $\frac{2500}{5} = 500$ cm³ per breath $\frac{8000}{5} = 1600$ cm³ per breath
      i.e. exercise has increased the depth of breathing.
   b. i. carbon dioxide
      ii. diffusion

Unit 12.1
LL   C: it is broken down to release energy
1. a. i. aerobic respiration
      ii. Towards the glass tube with the fruits in it
      iii. The fruits use up oxygen, and the carbon dioxide they release is absorbed by the soda lime. Therefore the volume of gas inside the tube falls.
      iv. $\frac{40}{5} = 8$. $8 \times 0.25 = 2$cm. Therefore the marker drop will be at 3.25 on the scale (i.e. current position is 5.25)
      v. suitable control would be the tube set up exactly the same but without the tomato fruits.
   b. Nitrogen gas does not allow respiration so that the fruits will not over-ripen and become 'spoiled'.

Unit 12.2
LL   C: lactic acid
1. a. i. $C_6H_{12}O_6 + 6O_2 \rightarrow 6CO_2 + 6H_2O$ + energy
      ii. Respiration is catalysed by enzymes, which are affected by temperature
   b. i. Pie chart B – most of the energy must be supplied by aerobic respiration as the athlete could not continue to run if anaerobic respiration was the energy supplier (much less efficient, and lactate is toxic).
      ii. Performance is reduced as the lactic acid stops impulses reaching the muscle cells.
      iii. The lactic acid is removed by diffusing into the blood
   c. The extra oxygen required to remove lactic acid accumulated during anaerobic exercise.

Unit 13.1
LL   D: excretion
1. a. The removal of toxins, the waste products of metabolism and substances in excess of requirements
   b. From top, and reading right to left: living cells   respiration   lungs   urea   liver   kidneys   absorption in gut   kidneys
   c. i. deamination (production of urea from excess amino acids)/storage of glycogen/removal of toxins such as alcohol
      ii. A – bile duct – bile; B – pancreatic duct – enzymes including amylase and protease; C – small intestine – food; D – hepatic portal vein – dissolved foods; E – hepatic artery – oxygenated blood; F – hepatic vein – deoxygenated blood

Answers

Unit 13.2
LL D: excretory
1. a. renal artery
   b. glomerulus
   c. The membrane is selective – pores are too small to allow passage of blood cells
   d. Mitochondria to supply energy for active transport/microvilli to provide increased surface area for absorption
   e. ureter

Unit 13.3
LL true
1. a. alkali/ammonia
   b. Optimum temperature for the activity of the enzyme urease
   c. i. protein and glucose
      ii. They are reabsorbed in the first coiled tubule of the nephron
   d. As a control, to show no result for these tests if only water is available to the urease

Unit 13.4
LL Because the patient's own immune system recognises the transplanted kidney as 'foreign' tissue, and attempts to reject it. Immunosuppressive drugs reduce the activity of the immune system.
1. a. To increase surface area for exchange
   b. It will increase the effective filtration pressure (equivalent to raising blood pressure)
   c. i. Higher concentration of urea, and sodium and chloride ions
      ii. ammonia   glucose   urea
      iii. lower/the same   the same   the same   lower   lower   lower

Unit 14.1
LL myelin; insulator; impulse/action potential
1. From the top of the table: A; C; E/F; E; D; B

Unit 14.2
LL rapid/involuntary/positive survival value
1. a. i. nose
      ii. They always have a survival value, e.g. clearing the eye of particles of dust
   b. i. 1 – receptor   2 – sensory neurone   3 – white matter
         4 – grey matter   5 – association neurone   6 – synapse
         7 – motor neurone
      ii. muscles – contract; glands – secrete (e.g. a hormone)

Unit 14.3
LL neurone; dendrites; axon
1. a. i. A molecule, produced in the synaptic vesicles and released across the synaptic cleft where it can bind with receptors on the postsynaptic membrane
      ii. A minute gap at a junction between adjacent neurones
   b. Operates as a valve – ensures impulses only travel in one direction
   c. Drugs affect the nervous system by affecting the binding of neurotransmitters to the receptors

Unit 14.4
LL B: stimulus
1. a. A Eye – to sight the ball; B (nose) – to smell; C touch receptor in skin – to feel the racquet in his hand; D semicircular canals in ear – (detects position)
   b. An electrical 'pulse, which travels as a wave of ion movements along the sensory neurones

Unit 14.5
LL B: rectum
1. a. From the top of the table: C; F; B; A; E; D.
   b. i. stimulus is bright light; receptor is retina; co-ordinator is central nervous system; effector is iris; response is reduce diameter of pupil;
      ii. Prevents bright light from damaging the retina

Unit 14.6
LL C: pancreas
1. a. Hormone: a chemical, produced by an endocrine organ, released into the bloodstream where it brings about a response at a target organ. Target organ: an organ which brings about a response to stimulation by a specific hormone
   b. Hormones: slower/longer-lasting/more general than nervous control

Unit 14.7
LL Homeostasis; negative
1. a. A: sensory neurone   B: hair shaft
   b. i. it is raised
      ii. Keeps a layer of air close to the skin, acting as a thermal insulator.
      iii. capillary constricted, position the same
   c. This is a control system in which a deviation from the norm sets up a series of changes which cancel out the deviation.

Unit 14.8
LL A: shivering
1. a. A: sweat duct/pore   B: dermis   C: branch of vein
   b. They act as insulation – to retain heat within the body
   c. Optimum temperature for enzymes/cell membranes do not break down/solutes remain dissolved in plasma

Unit 14.9
LL AUXIN; GRAVITROPISM; PHOTOTROPISM
1. a. i. Aauxin has moved downwards and to the non-illuminated side of the shoot.
      ii. Auxin moves to non-illuminated side – causes enlargement of cells on this side – shoot therefore bends towards the light.
      iii. response – (positive) phototropism; benefit – moves leaves into the optimum position for photosynthesis
   b. i. Coordinating formation of fruit for easier harvesting
      ii. Acting as a weedkiller in growing crops

Unit 15.1
LL False
1. a. Heroin – feeling of calm and rest; nicotine – stimulates the heart rate; oestrogen – stimulates ovulation; penicillin – protects against syphilis; testosterone – increases growth rate in muscle
   b. An inability of the body to function normally in the absence of a drug
   c. Antibiotic resistance is the property of some bacteria which cannot be killed or inhibited by a particular antibiotic. It might be dangerous in the case of infection, as the antibiotic may not be effective and the infected person might not recover.

Unit 15.2
LL depressant/narcotic; brain; addiction; withdrawal
1. a. i. Heroin stimulates pleasure and calmness receptors so that the pain is 'hidden'
      ii. The level of a drug needed to cause an effect is increased, usually because additional receptors for the natural neurotransmitter is increased
   b. Loss of work ethic/dishonesty in order to obtain money for drug purchase/loss of social awareness

Unit 15.3
LL EPO; oxygen; muscles
1. a. From top of table: C; A; B; E/F
   b. Alcohol transported in maternal blood/diffuses across placenta/enters fetal bloodstream

Unit 15.4
LL C: cholera
1. a. i. Effect of age on the likelihood of COPD
      ii. Chronic obstructive pulmonary disorder
      iii. No – there is no information about smoking, only about the age of sufferers from COPD

Answers

  b. Vital capacity would fall. This is because smoking damages the alveoli so that they become less elastic. As a result they cannot stretch so the vital capacity is reduced.

Unit 16.1
LL B: the offspring are genetically identical
1. a. mitosis in the asexual diagram, meiosis in the sexual diagram.
   b. for asexual   nn   and   nn;   for sexual   n   and   n
   c. advantage: rapid/progeny well adapted to same environment as parents
      disadvantage: no variation possible, so great risk if 'unknown' infection attacks population

Unit 16.2
LL C: anther
1. Transfer of pollen from anther to the stigma.
   Z: petal;   W: anther;   Y: stigma;   X: ovary
   Petals: small and dull-coloured;   large and brightly coloured; attraction of insects
   Anthers: hang out into the wind;   kept deep inside flower; anthers can release pollen into the wind
   Pollen: very light;   heavy and produced in smaller quantities; light pollen can float in the wind
   Stigma: branched;   deep within flower, strong;   branched to catch wind-blown pollen/strong to avoid damage from visiting insect

Unit 16.3
LL D: transfer of pollen from stamen to stigma
1. a. i. Mark P on tip of stigma
      ii. stigma
   b. i. Show entry via the micropyle
      ii. Male nucleus fuses with female nucleus to form a diploid zygote. Permits variation by random combination of genetic material from parent plants.

Unit 16.4
LL Embryo; starch; testa
1. a. From the top of the column: E; C; B; A; D; F
   b. Show tube growth down style and entry to ovule via the micropyle
   c. i. plumule and radicle
      ii. starch
      iii. Iodine solution has changed from straw-brown to blue-black

Unit 16.5
LL D: they contain the haploid number of chromosomes
1. a. A: sperm duct/vas deferens;   B: urethra;   C: testis;
      D: scrotal sac/scrotum;   E: penis;   F: prostate gland
   b. i. –v. in order; C; F; C; B; A
   c. E – A – D – B – F – C

Unit 16.6
LL B: ovary
1. a. B
   b. A
   c. E
   d. C
   e. D

Unit 16.7
LL A: uterus
1. a. From top of diagram: oviduct; ovary; uterus; vagina
   b. i. in oviduct/uterus
      ii. zygote
   c. From top of column: fertilisation; conception; copulation; AID; implantation; development

Unit 16.8
LL A: gestation period
1. a. i. amniotic cavity – contains amniotic fluid which acts as a shock absorber; uterus wall – muscular and so is protection against physical damage
      ii. from end of umbilical cord to wall of uterus
   b. Glucose: from mother to fetus; haemoglobin: no movement across placenta; nicotine: from mother to fetus; amino acids – from mother to fetus; carbon dioxide: from fetus to mother; alcohol: from mother to fetus; urea: from fetus to mother

Unit 16.9
LL A: sugar
1. a. $\frac{2400}{8800} \times 100 = 27\%$
   b. More calcium and vitamin D for rapid growth of developing bones; less iron because the infant has less need to replace haemoglobin as it has a lower blood volume
   c. $\frac{2400}{2750} = 872$ g
   d. Cow's milk has too much protein, not enough iron or vitamin D.
   e. Vitamin D is synthesised in the skin when the skin is exposed to sunlight.

Unit 16.10
LL puberty; sex hormones; secondary sexual
1. a. i. pituitary gland
      ii. testosterone
   b. breasts – for production of milk; hips – to accommodate the growing fetus in the uterus
   c. menopause

Unit 16.11
LL 56
1. a. i. 8
      ii. 5th July
      iii. no egg cell is available
   b. i. progesterone
      ii. oestrogen

Unit 16.12
LL condoms/barrier methods; HIV/gonorrhoea
1. a. Prevents conception/prevents transmission of infected body fluids
   b. i. From the top of the column: 5 – 3 – 4 – 2 – 6 – 1
      ii. Combination of progesterone and oestrogen reduces the chance of ovulation (feedback inhibition). No ova available means no possibility of fertilisation/conception.
      iii. $1000 \times \frac{1}{20} = 50$

Unit 16.13
LL IVF = in vitro fertilisation; AI = artificial insemination
1. a. LH and oestrogen would combine to increase the possibility of ovulation
   b. FSH increases the production of ova
   c. Feedback inhibition of LH and FSH

Unit 16.14
LL C: white blood cells
1. a. i. bacterium
      ii. painful urination
      iii. condom
   b. i. Reduces the production of defensive white blood cells
      ii. AIDS is viral, and so is not affected by antibiotics, which can control the bacterium causing gonorrhoea.

Unit 17.1
LL D: gene
1. a. GCCTATG
   b. i. messenger RNA
      ii. ribosome
   c. amino acid

Answers

d. i. protein
    ii. haemoglobin – antibodies – can recognise and bind to a neurotransmitter – enzyme which breaks down fats to fatty acids and glycerol – keratin

Unit 17.2

LL ribosomes; amino acids; genes
1. a. amylase
    b. Advantages: protein can be made more cheaply/in can be purer/it can be produced when required
        Concerns: modified organisms might escape into the environment/superweeds could be created if modified plants pollinate wild plants/genetically-modified products could be expensive if the organisms are patented by drug companies.

Unit 17.3

LL C: two cells with 46 chromosomes each
1. a. Male nerve cell: 46 XY; female white blood cell 46 XX; sperm cell: 23 X or Y; egg cell: 23 X; red blood cell 0 none
    b. mitosis
2. a. 4
    b. mitosis
    c. bone marrow/liver/skin/apical meristem of plant

Unit 17.4

LL B: red blood cell (no chromosomes) and sperm cell (haploid)
1. a. One large and one small in each of the daughter cells
    b. i. ovary/testis
        ii. anther/ovule
    c. To reduce the number of chromosomes to half so that at fertilisation the 'normal' (diploid) number is restored

Unit 17.5

LL A: have brown eyes but different genotypes
1. gene   meiosis   haploid   fertilisation   diploid   recessive   heterozygous
2. genotype – the set of alleles present in an organism; homozygous – having two identical alleles; dominant – an allele that is always expressed if it is present; heterozygous – having two alternative alleles; recessive – allele that is only expressed in a homozygous individual; chromosome – a thread-like structure of DNA…; allele – one alternative form of a gene; phenotype – the observable features of an organism

Unit 17.6

LL C: the child could receive the recessive allele from both parents
1. a. i. gender/eye colour/blood group
        ii. an alternative form of a gene
        iii. Show a number of genes lined up along a thread-like chromosome
    b. i. 9
        ii. could be Rr or RR
        iii. $\frac{3}{4}$ or 75%

Unit 17.7

LL A: AO and BO
1. a. i. Andrew and John
        ii. blood group – a discontinuous variation, so affected by genes only.
        iii. nutrition – David might eat more high-calorie foods, or take less exercise
    b. codominance

Unit 17.8

LL D: a girl who is a carrier for colour blindness
1. a. female
    b. An extra 21st chromosome
2. a. 2 has 2 X chromosomes, 3 has two X chromosomes 5 has one X and one Y chromosome
    b. 2 could have H and h, 3 could have h and h, 5 could be h
    c. female – female – female – male – male

Unit 18.1

LL phenotype; genotype; environment
1. continuous variation – a form of variation with many intermediate forms between the extremes; gene – a section of DNA responsible for an inherited characteristic; discontinuous variation – a form of variation with clear-cut differences between groups; phenotype – the observable features of an organism; height in humans – one example of continuous variation; environment – this factor, in addition to genotype, can affect phenotype; nutrients – one possible form of environmental influence on variation; blood group – one example of discontinuous variation.

Unit 18.2

LL C: excess vitamin C
1. a. i. $Hb_S Hb_A$
        ii. Tom, since both parents are heterozygous (carriers of the abnormal allele)
    b. This allele offers some protection in areas where malaria is common. The malarial parasite cannot reproduce so easily inside the red blood cells of a sickle cell individual.
2. a. i. Organisms with so many common features that they may interbreed and produce fertile offspring.
        ii. A change in the type or quantity of DNA in an individual.
        iii. A section of DNA responsible for coding for a single protein/characteristic
    b. Radiation/carcinogenic chemicals such as tar in cigarette smoke

Unit 18.3

LL Adaptation; genotype
1. a. Artificial – artificial – natural – natural – natural
    b. i. By selecting individual animals with a high milk yield, breeding them, then selecting from their offspring cows that have a high milk yield, then breeding them, and so on.
        ii. resistance to disease/ability to withstand difficult environmental conditions such as low temperatures

Unit 18.4

LL B: is well camouflaged from its natural predators
1. a. B – E – C – A – D
    b. Heron – feeds by spearing fish and frogs; hawk – captures Florida rabbits and other mammals; spoonbill – filters algae and other small organisms from the water; finch – feeds on nuts and other hard fruits; warbler – feeds by catching small insects

Unit 18.5

LL species; artificial selection/selective breeding
1. a. brussels sprouts – bud; broccoli – flower
    b. i. They have been produced by mitosis, so there is very little possibility of genetic variation.
        ii. nitrate – the ion needed for protein synthesis (growth); plants in closed environment – temperature and humidity can be controlled so that cuttings do not dry out

Unit 19.1

LL B: solar energy from the Sun
1. a. i. the Sun
        ii. feeding
        iii. Some is reflected/the wrong wavelength/does not fall on leaves
        iv. respiration – energy is lost here as heat
    b. i. A: $\frac{15000}{90000} \times \frac{100}{1} = 16.6\%$ B: $\frac{2000}{15000} \times \frac{100}{1} = 13.3\%$
        ii. Each stage in the food chain allows a loss of energy as heat. Fewer steps, as in eating a vegetarian diet, mean that more energy is transferred with this heat loss.

Answers

Unit 19.2
LL mass of individuals; number of individuals
1. Food chain – the transfer of energy from one organism to the next, beginning with a producer; food web – a network of interconnected food chains; producer – an organism that makes its own organic nutrients, usually through photosynthesis; consumer – an organism that gets its energy by feeding on other organisms; herbivore – an animal that gets its energy from eating plants; carnivore – an animal that gets its energy by eating other animals; decomposer – an organism that gets its energy from dead or waste material
2. Pyramid should have broad base, of algae, then gradually narrowing through water fleas, smelt and to the pointed peak at kingfisher

Unit 19.3
LL A: kilojoules
1. a. There are fewer steps in the food chain leading to grains and seeds, so fewer opportunities for loss of energy
   b. Less of the food supplied to the cattle is used to generate heat to maintain their body temperature, so more can be used for protein synthesis/growth.

Unit 19.4
LL photosynthesis; respiration
1. a. From left to right and top to bottom: photosynthesis; respiration; feeding; decay
   b. It would raise the carbon dioxide concentration in the air

Unit 19.5
LL Nitrate; nitrogen gas; denitrifying bacteria
1. a. i. B – feeding   E – denitrification
      ii. Decomposition/decay by bacteria and fungi
   b. i. Ammonium concentration would rise/nitrite and nitrate levels would fall/fewer dead leaves would be decomposed
      ii. Ticks in few scavenging insects/ regular turning of the heap to add air

Unit 19.6
LL Lag – exponential – stationary – decline
1. a. higher death rate/less food/lower birth rate/ emigration
   b. vaccination/immunisation programmes; use of antibiotics; improved surgical techniques; fewer deaths at childbirth
   c. use of antibiotics/antiseptics; thorough cooking to reduce food poisoning bacteria

Unit 19.7
LL agricultural; industrial; medical
1. a. The percentage of the population made up of the different age groups
   b. 1: food availability – poor diet can lead to high death rate in younger age groups; 2: hygiene – food poisoning can reduce populations, especially in the very young and the elderly; 3: medical provision – immunisation programmes reduce death rates in lower age groups

Unit 20.1
LL D: plasmid
1. a. From the top of the table: F – F – T – T – F – T – T – T – T – T – T – F
   b. Typical bacterium is one millionth of a metre wide/bacteria are larger than viruses/many bacteria are not harmful, they are useful/ bacteria do not have a nucleus – they have 'naked' DNA

Unit 20.2
LL pectinase
1. a. from the top: milk containing lactose   lactase   lactose-free milk
   b. Immobilisation – fixed onto an inert substance
      Importance – allows re-use of enzyme/provides product free of enzyme
   c. if present , use answers from CB answers.

Unit 20.3
LL B: fungi
1. a. (6–) 7 days: this represents the maximum yield of penicillin – any time longer produces less penicillin but would be expensive in energy and raw materials
   b. Some bacteria in the population have a natural resistance – the non-resistant bacteria are killed by the antibiotic – the resistant bacteria now multiply and produce a resistant population
   c. pectinase – protease – removes milk sugar from milk

Unit 20.4
LL B: insulin
1. a. Gene – a section of DNA coding for a protein; plasmid – a small circle of DNA in a bacterial cell; vector – a structure which can carry a gene into another cell; ligase – an enzyme that can splice one gene into another section of DNA; restriction – an enzyme that can cut a specific gene from a chromosome; sticky ends – pieces of single stranded DNA left exposed after a gene is cut from a chromosome; fermenter – a vessel in which engineered bacteria can produce a valuable product under optimum conditions
   b. Insulin – control of blood sugar level; factor 8 – one step in blood clotting; pectinase – 'clears' fruit juices by breaking up clumps of plant tissue; human growth hormone – can increase growth rate in humans of short stature
   c. The virus might not be completely inactivated, and thus may infect a patient with hepatitis

Unit 21.1
LL D: milk
1. a. i. beans
      ii. rice
      iii. vitamins and minerals
   b. drought/loss of nutrients from soil/displacement of populations by war

Unit 21.2
LL nesting sites; soil
1. loss of habitat for animals / possible soil erosion / changes to water cycle / possible loss of valuable medicinal plants
2. a. i. Marsh has been drained/hedges have been removed/there are more buildings in 2003.
      ii. Marsh animals would be reduced in number, e.g. frogs need water in which to lay their eggs; hedges removal would mean fewer nesting sites for birds; hedge removal would eliminate certain food plants for insects such as butterflies.
   b. Loss of nesting sites/soil erosion/loss of food plants for some species/less removal of carbon dioxide from the atmosphere/ disturbance of water cycle.

Unit 21.3
LL A: sulfur dioxide and nitrogen oxides
1. a. The gases form a layer around the Earth which allows radiation in but does not allow thermal radiation out – the surface of the Earth therefore becomes warmer
   b. 1. These renewable energy resources do not produce greenhouse gases
      2. Insulation means less heat loss so less combustion of fossil fuels for heating
      3. Reforestation increases the number of plants which can remove carbon dioxide from the atmosphere during photosynthesis
2. Removal of habitat for road construction / disturbance of wildlife by noise and presence of humans / pollution from vehicle exhausts

Unit 21.4
LL A: are washed into rivers and pollute the water
1. a. C – E – B – D – A
   b. Run-off of fertilisers from nearby farmland
   c. 1. Bacteria respire aerobically, using dead plants as food source
      2. Aerobic respiration by bacteria reduces oxygen concentration in the water
      3. Fish and larger invertebrates die as they are short of oxygen

Answers

Unit 21.5

LL carbon dioxide; methane

1. a. carbon dioxide + water ⟶ glucose + oxygen: this process is driven by light energy
   b. The biocoil must be transparent to allow light to penetrate to the algae
   c. The algae are removed to use as a fuel – they must be replaced quickly if the process is to be economical

Unit 21.6

LL reduces; leaching

1. a. Increase defoliation/increase leaching of nutrients from the soil
   b. car exhausts
   c. scrubbers on chimneys/burn less fossil fuels/catalytic converters on cars
   d. damage to the lining of the lungs – breathing difficulties

Unit 21.7

LL C: legumes

1. a. More shoals of fish are located, and more are caught by the fine mesh. Few fish escape to breed, so the population decreases. Loss of one species will affect the numbers of other species in the food web for this habitat.
   b. Increase size of mesh/restrict areas in which fishing can take place/restrict times of year in which fishing can take place/limit the size of the catch for each fishing vessel

Unit 21.8

LL chlorine

1. a. In the absence of oxygen
   b. i. carbon dioxide
      ii. As dry ice for cooling foods in cold stores
   c. To avoid growth of potentially harmful microbes in the lake – the lake water might be consumed by humans
   d. As a fertiliser – it has high concentrations of nitrate and phosphate
   e. Disinfectants could kill the beneficial bacteria which decompose the sewage

Unit 21.9

LL Habitat; reproduce

1. Management of the environment to maintain biodiversity
2. a. i. Environmental factor: removal of bamboo forests – bamboo shoots are the basic food of the Giant Panda
      Biology: pandas reproduce very slowly – number of births does not replace number of dying pandas, so population falls
      ii. Saving the Panda (a very attractive species which many people will help to conserve) also protects other species in the same habitat
   b. i. To find the population size at the start of the management plan
      ii. Many examples possible, e.g. butterfly – with nets, beetles – with pitfall traps, plants – with quadrats

Unit 21.10

LL biodiversity; humans

1. a. A process which does not affect the environment for future generations
   b. Less competition for light/for nutrients/for water
   c. Limited food variety for animals/any pest or disease can affect all of the plants
   d. Any of them might be a benefit. Best might be fast growth (so product is available quickly), resistance to disease (so few plants are lost to disease) and good growth on poor soil (so few fertilisers are necessary).

Language focus
Unit 22
Matching pairs

Food chain – the transfer of energy from one organism to another; producer – an organism that makes its own organic nutrients, usually using energy from sunlight: carnivore – an animal that gets its energy by eating other animals; trophic level – the position of an organism in a food chain; food web – a network of interconnected food chains; consumer – an organism that gets its energy by feeding on other organisms; herbivore – an animal that gets its energy by eating plants; decomposer – an organism that gets its energy from dead or waste organic material; community – all of the populations of different species in an ecosystem; population – a group of organism of one species, living in the same area at the same time; ecosystem – a unit made up of the community of organisms interacting with their non-living environment

Word scrambles

SEXUAL REPRODUCTION; GAMETES; ZYGOTE; FERTILISATION; PUBERTY; SEX HORMONES; TESTES; OVARIES; VAGINA; PENIS; CONCEPTION; COPULATION; EMBRYO; MENSTRUATION; GROWTH; DEVELOPMENT; PLACENTA; GESTATION PERIOD; UMBILICAL CORD; AMNIOTIC FLUID

Crossword: DNA

Across: 1 – sugar phosphate   3 – double helix   5 – chromosome
6 – gene   7 – Watson and Crick   8 – carrier protein   13 – nucleus
14 – deoxyribonucleic acid   15 – nucleotide   16 base   17 – franklin
Down: 2 – haemoglobin   4 – base pairing   9 – replication
10 – protein   11 – mitosis   12 – keratin

Crossword: Inheritance

Across: 5 – genetic engineering   7 – cystic fibrosis   8 – plasmid
9 – virus   10 – chimaera   11 – hormones   12 – gene therapy
14 – haemophilia   16 – propagation   17 – fermenter   18 – vegetative
20 – insulin   23 – cellulase   24 – enzymes   26 – vector   27 – clone
28 – protoplast
Down: selective breeding   2 – DNA   3 – meristem   4 – milk
6 – bacteria   13 – protein   15 – amniocentesis   19 – nucleus
21 – humidity   22 – gene   25 – dolly

Maths for Biology
Unit 26.1

1. a. 10
   b. 1000
2. a. A – cell surface membrane, B – cytoplasm, C – nucleus
   b. nucleus
   c. cellulose cell wall/chloroplast/permanent vacuole
   d. 3 mm = 3000 μm   Mag = $\frac{3000}{25}$ = ×120

Unit 26.2

1. a. Age group on x axis, frequency on y axis
   b. Increasing age leads to increasing frequency of multiple births (give some numerical examples to support your answer)

Unit 26.3

1. a. temperature on x axis, time taken on y axis
   b. 35°C
   c. these are fixed variables
   d. Enzymes is denatured – loss of active site so cannot have binding of substrate to enzyme
   e. pH/enzyme concentration/substrate concentration

Unit 26.4

1. a. 0.63 − 0.42 = 0.21. $\frac{0.21}{0.42} \times \frac{100}{1} = 50\%$
   b. Releasing carbon dioxide from cylinders of compressed gas gives good control of carbon dioxide concentration

Answers

Unit 26.5

1. a. Using a pH meter on a sample of saliva collected (e.g. by chewing a rubber band)
   b. pH affects the number of fillings
   c. hypothesis supported; the increased pH shows a reduction in the number of fillings

Exam-style questions
Unit 27.0

1. C: cell wall
2. Organism B
3. D
4. A
5. C
6. B
7. B
8. D
9. C
10. C
11. a. Correct position above and to the right of the stomach
    b. peristalsis
    c. From the top of the column: A – A – C – E – F
12. a. i. A: respiration; B: photosynthesis C: feeding D: respiration.
       ii. decay/decomposition – a process in which organisms secrete enzymes which break down complex organic compounds such as proteins into simpler compounds such as amino acids/ammonium nitrate
    b. i. In compost heap respiration is taking place. This releases energy, some of which is lost as heat.
       ii. This allows oxygen to penetrate the compost – oxygen is needed for aerobic respiration
13. a. i. large surface area/close to blood supply/moist lining
       ii. Allows stretching during the inhalation/exhalation cycle
    b. Blood at A has more carbon dioxide and less oxygen than blood at B
    c. Makes lung tissue less elastic/breaks down walls of alveoli/inactivates cilia in airways/may trigger development of cancer cells
14. a. i. A change in the quantity or type of DNA
       ii. cystic fibrosis/Huntington's disease/albinism/sickle cell anaemia
    b. i. A recessive allele is only expressed in a homozygous individual
       ii. A female affected with phenylketonuria
       iii. individual 4: Pp; individual 6: PP or Pp
       iv. Individual 10 must be heterozygous, i.e. Pp. Thus child has 1 in 4 chance of being homozygous, i.e. showing phenylketonuria

Data sheets

You will find that the practical paper or the alternative to practical will be much more straightforward if you recognise certain standard pieces of equipment, and understand what they are used for.

Apparatus and materials

Safety equipment appropriate to the work being planned, but at least including eye protection such as safety spectacles or goggles.

3D image	Name and function	Diagram
	Watch glass: used for collection and evaporating liquids with no heat, and for immersing biological specimens in a liquid.	
	Filter funnel: used to separate solids from liquids, using a filter paper.	
	Measuring cylinder: used for measuring the volume of liquids.	
	Thermometer: used to measure temperature.	

The other very important measuring device in the laboratory is a balance (weighing machine).

	Spatula: used for handling solid chemicals; for example, when adding a solid to a liquid.	
	Pipette: used to measure and transfer small volumes of liquids.	
	Stand, boss, and clamp: used to support the apparatus in place. This reduces the risk of dangerous spills. This is not generally drawn. If the clamp is merely to support a piece of apparatus, it is usually represented by two crosses as shown.	
	Bunsen burner: used to heat the contents of other apparatus (e.g. a liquid in a test tube) or for **directly** heating solids.	HEAT
	Tripod: used to support apparatus above a Bunsen burner. **The Bunsen burner, tripod, and gauze are the most common way of heating materials in school science laboratories.**	
	Gauze: used to spread out the heat from a Bunsen burner and to support the apparatus on a tripod.	
	Test tube and boiling tube: used for heating solids and liquids. They are also used to hold chemicals while other substances are added and mixed. They need to be put safely in a test tube rack.	
	Evaporating dish: used to collect and evaporate liquids with or without heating.	
	Beaker: used for mixing solutions and for heating liquids.	

Data sheets

Chemical reagents

In accordance with the COSHH (Control of Substances Hazardous to Health) regulations operative in the UK, a hazard appraisal of the list has been carried out. The following codes are used where relevant.

C = corrosive substance

F = highly flammable substance

H = harmful or irritating substance

O = oxidising substance

T = toxic substance

Table of hazard symbols

Symbol	Description	Examples
	Oxidising These substances provide oxygen which allows other materials to burn more fiercely.	Bleach, sodium chlorate, potassium nitrate
	Highly flammable These substances easily catch fire.	Ethanol, petrol, acetone
	Toxic These substances can cause death. They may have their effects when swallowed or breathed in or absorbed through the skin.	Mercury, copper sulfate

Symbol	Description	Examples
	Harmful These substances are similar to toxic substances but less dangerous.	Dilute acids and alkalis
	Corrosive These substances attack and destroy living tissues, including eyes and skin.	Concentrated acids and akalis
	Irritant These substances are not corrosive but can cause reddening or blistering of the skin.	Ammonia, dilute acids and alkalis

Data sheets

Reagent	Use in biology
hydrogencarbonate indicator (bicarbonate indicator)	Detects changes in carbon dioxide concentration, for example in exhaled air following respiration
✖ iodine in potassium iodide solution (iodine solution)	Detection of starch
✖ Benedict's solution (or an alternative such as Fehling's)	Detection of a reducing sugar, such as glucose
⚠ biuret reagent(s) (sodium or potassium hydroxide solution and copper sulfate solution)	Detection of protein
🔥 ethanol/methylated spirit	For dissolving lipids in testing for the presence of lipids. Also used to remove chlorophyll from leaves during starch testing
cobalt chloride paper	Detects changes in water content
pH indicator paper or Universal Indicator solution or pH probes	Detects changes in pH during reactions such as digestion of fats to fatty acids and glycerol
litmus paper	Qualitative detection of pH
glucose	Change in water potential of solutions
sodium chloride	Change in water potential of solutions
aluminium foil or black paper	Foil can be used as a heat reflector: black paper as a light absorber
a source of distilled or deionised water	Change in water potential of solutions. Also used in making up mineral nutrient solutions.
eosin/red ink	To follow the pathway of water absorbed by plants
limewater	A liquid absorbent for carbon dioxide, for example in exhaled air
✖ methylene blue	A stain for animal cells
⚠ potassium hydroxide	Removes carbon dioxide from the atmosphere, for example during experiments on conditions for photosythesis
sodium hydrogencarbonate (sodium bicarbonate)	Very mild alkali: use for adjusting pH of solutions for enzyme activity
vaseline/petroleum jelly (or similar)	Blocks pores such as stomata, and so prevents water loss